Praise for

BREAKING TRAIL

"It's hard to resist the appeal of an author who spent three months trekking across the European Alps from Yugoslavia to France with a newborn daughter strapped to her back. But the challenges that Arlene Blum has tackled during her life as a hiker, adventurer, and biochemist are precisely what render *Breaking Trail* a book too engrossing to put down."
—*The Seattle Post-Intelligencer*

"Inspiring . . . [Blum] is challenged often on the road to becoming a distinguished mountaineer and chemist, but responds with courage and determination."
—*Calgary Herald*

"Compelling . . . Blum exudes possibility."
—*Los Angeles Times*

"Personal and disarmingly honest . . . [Blum] simply tells her nourishing and deserving story, quietly reminding us that a woman's place is indeed on top."
—*The New York Times Book Review*

"A magnificent and compelling story. Blum leads the reader into the beautiful, exciting, and terrifying world of mountain climbing. Her writing soars. She conveys the drama and mind-set that it takes to go to places where few have ventured. Her inspiring story is as much about leadership as it is about living life fully and overcoming obstacles to reach one's goals. It's a great book."
—Lynne Cox, author of *Swimming to Antarctica*

"*Breaking Trail* is the best leadership autobiography I have read. It's not only a case study of decision-making under extreme, life-and-death situations but also a story of one person's adventurous path through the life lessons that define character. Required reading for anyone who aspires to lead others to greatness."

—James Kouzes, coauthor of *The Leadership Challenge*

"This is an engaging, well-written adventure that also serves as a social history of women's roles. It should be required reading for young women of today."

—*Booklist*

"Her legacy as a pioneer lends Blum's story a historical resonance."

—*The Washington Post*

"A classic read that parallels the history and plight of the American woman."

—*The Denver Post*

"Eye-opening . . . Blum is a very likeable and humble guide, with an optimistic spirit."

—*The Jewish Week*

"*Breaking Trail* is one of those rare books that manage to be both intimate and universal. I found myself inspired—and educated—by Arlene Blum's courage in the face of continual obstacles, many far more daunting than a crevasse or icefall. More than a great read about climbing, it's a terrific allegory for the struggle that so many brilliant women have faced on their paths to the summit."

—Jeff Greenwald, author of *The Size of the World* and *Shopping for Buddhas*

"This book is a must-read for those who wish to reach the highest level of personal fulfillment."

—Helen Thayer, author of *Polar Dream*

ARLENE BLUM

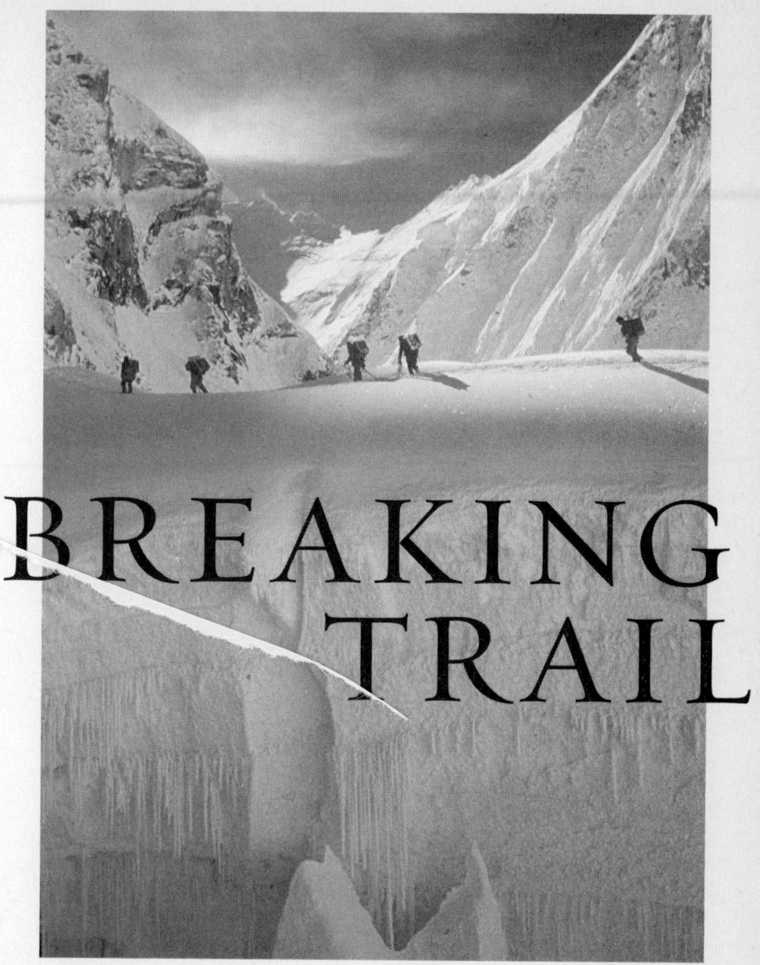

BREAKING TRAIL

A Climbing Life

A HARVEST BOOK | HARCOURT, INC.
Orlando Austin New York San Diego Toronto London

www.HarcourtBooks.com

First published by Scribner in 2005.

Library of Congress Cataloging-in-Publication Data
Blum, Arlene, 1945–
Breaking trail: a climbing life/Arlene Blum.—1st Harvest ed.
p. cm.—(A Harvest book)
Includes index.
1. Blum, Arlene, 1945– 2. Mountaineers—United States—Biography. 3. Women mountaineers—United States—Biography.
I. Title.
GV199.92.B595A33 2007
796.522092—dc22
[B] 2006031752
ISBN 978-0-15-603116-5

Text set in Requiem
Designed by Linda Lockowitz

Printed in the United States of America
First Harvest edition 2007
K J I H G F E D C B A

Photographs by Arlene Blum or from her collection unless otherwise noted.

To my beloved daughter, Annalise,
and all who break their own trails

Contents ▲

Foreword ▲

by Sir Chris Bonington

Arlene Blum has led a remarkable life. From unlikely beginnings, she became a leader in the breakthrough of women into mountaineering on the world's highest peaks. Hers is a compelling narrative on many levels—it is a warm and intimate memoir, an important account of the development of women's climbing, and a dramatic adventure story.

Climbers' autobiographies and biographies all too often focus on their ascents to the exclusion of anything personal. Arlene, by contrast, is courageously open about her private life. As well as riveting accounts of her expeditions, she shares with the reader her childhood travails, passage through university, discovery of climbing, relationships on and off the hill, struggles with chauvinistic colleagues, and work as a distinguished research scientist.

These all come together in a beautifully crafted book describing a full and fascinating life. I found myself empathizing with her, for I also was brought up by a single parent—my mother—helped by my grandmother, with conflict between them. I, too, was initially awkward in forming relationships with the opposite sex, but was immensely fortunate in finding the woman of my life when I was twenty-seven.

I also share with Arlene the rewards and stresses of leading high-altitude mountaineering expeditions. I have had the same doubts about my ability to lead, hassles with individualistic fellow climbers, and dilemmas in balancing my climbing and personal life. And I, too, have lost all too many close friends.

The hurdles I faced, however, were much lower than those that confronted Arlene. I'm an Anglo-Saxon male, and this makes things a great deal easier. Arlene started her climbing career at a time when women were perceived as dutiful seconds who held the rope, made the tea, and did what they were told. In trying to climb with men on an equal basis, Arlene suffered rejections both from individuals and the climbing establishment. Yet she persevered, establishing firsts for women and also planning and carrying out a series of unique adventures, including her inimitable Endless Winter.

Her leadership of the American Women's Himalayan Expedition was an exceptional achievement. Annapurna I is one of the most serious of the 8,000-meter peaks and a desperately dangerous mountain. Arlene set aside trying to reach the summit herself to support the efforts of her teammates. When I led my expedition to Annapurna South Face in 1970, I similarly had to cope with the diverse aspirations of my team and devastating tragedy after success had been attained.

This memoir is gripping throughout and extraordinarily rewarding at the end, when Arlene's childhood, climbing, and scientific careers come together in a surprising and satisfying manner. I heartily recommend *Breaking Trail* to both men and women climbers (and to armchair mountaineers) as well as to anyone who faces uphill struggles. We learn from Arlene's story that with conviction and persistence, we can achieve our most challenging and improbable goals.

Nether Row, Cumbria
United Kingdom
March 2005

Introduction ▲

"The top! We made it!" The six of us cheered and hugged. We were the first team of women to reach the arctic summit of Denali, the highest mountain in North America. Many people believed that women lacked the physical strength and technical skill to climb the toughest mountains. Our ascent demonstrated to the world and ourselves what women could achieve.

All around us, the peaks of the Alaska Range extended to the horizon like frozen waves on a turbulent sea. Looking 8,000 feet down the vertical south face, we saw a thickening blanket of dark clouds. Although a storm raged below, it was warm and windless here at 20,320 feet.

We shared a quick lunch, but there was little time for celebration. Grace, our leader, was ill and getting worse by the moment. Moving slowly all day, she had insisted on continuing up and had barely made it to the top. Now she lay slumped in the snow, pallid and still. We had to get her to a lower elevation—and fast.

Margaret and Faye led Grace down the summit ridge on a short rope. Then Dana and I each took one of Grace's arms and supported her across a plateau the length of several football fields. She staggered between us in a stupor, her weight dragging us down into the sun-softened snow. "One step at a time," I encouraged her. "You've got to keep moving." We managed to get her across the flat area and back up an easy rise to a ridge at 19,600 feet, where she fell onto the snow, retching.

"Try to drink a little." Faye held her water bottle to Grace's chapped lips.

"Stop bothering me," Grace moaned. "I'm finished."

On the summit of Denali (Mt. McKinley), July 6, 1970.
(From left, standing) Dana Isherwood, Margaret Young, and Margaret Clark.
Faye Kerr and Grace Hoeman are in front. I'm the photographer.

I grasped her hand and tried to pull her up. "We've got to keep moving, Grace."

"Go away." She jerked her hand back and sank down, murmuring, "I'm going to die. Leave me here in peace." Her eyes closed and she drifted into unconsciousness.

I was terrified. As the deputy leader, I needed to take charge. Our camp was three thousand vertical feet below, it was seven in the evening, and an Arctic blizzard could engulf us at any moment. Exhausted from our long ascent, we had to get Grace down the mountain or stay up here with her. Both options seemed impossible.

Our team needed a strong leader and a sound plan of action. And so, at age twenty-five, on the frigid apex of North America, with storm clouds massed below and the specter of disaster in Grace's inert body, I reluctantly became an expedition leader.

Since that life-changing day on Denali, I have frequently asked myself why I spend my time and money to sentence myself to lack of oxygen, fierce weather, hard physical labor, and possible death. Each time, I resolve that on my next vacation, I'm going to the beach. But invariably, I find myself once again breaking trail through deep heavy snow or trying to sleep on a narrow, icy ledge.

To understand why I love climbing distant mountains, I decided to look close to home, at my upbringing. As a child who was not allowed to cross the street, literally or figuratively, I learned how to find my way through or around most barriers. Like a compressed spring, I was catapulted by my narrow, overprotected early years into the heights. Reliving my childhood while writing this book, I have discovered surprising solutions to some family mysteries as well as unexpected roots of my ability to lead mountaineering expeditions, solve scientific problems, and turn far-fetched visions into reality.

This memoir consists of short childhood vignettes, longer stories of my mountain adventures, and snapshots of my career as a scientist. Intertwined, these three strands provide insights into how and why I left the flatlands of the Midwestern United States for the steep slopes of mountains around the world to find a fulfilling life as a mountaineer, a scientist, and a mother.

ARLENE BLUM
Berkeley, California
April 2005
www.arleneblum.com

Chapter 1 ▲

Under the Porch

Davenport, Iowa, August 1949

The sun blazes relentlessly on me, so I grab my doll and climb into a cool, dark space under the back porch. I prop my doll up for a tea party. Between us I spread a lace handkerchief and lay out the blue glass doll dishes Mommy gave me for my fourth birthday. My aunts Ruth and Shirley are sitting on the porch above and their muffled words drift down. Lulled by the warm day, the drone of my aunts' voices, and the sweet smell of the rosebushes that surround our house, I begin to doze.

I hear my name, startle awake, and listen.

"With parents like that."

I strain to hear their words. There's something about my father, Germany . . . And then:

"Arlene . . . that child will amount to no good . . ."

Tears begin to blur my eyes. I curl up on the ground, hug my knees, and shake with silent sobs. I hate my aunt's words. I hate my aunt. I hate myself. But she is wrong. I'll show her. I'll show them all.

A Slide down Mt. Adams ▲

1964

"Can you keep going?" John handed me his water bottle.

"I'll try," I gasped, taking a sip. We continued upward in the dark, my loud breathing synchronized with the rhythmic tap of our ice axes on the rocky ground, the snow, the ice.

It was September 1964 and we were climbing Mt. Adams, a stately 12,276-foot volcano in southern Washington near Portland, Oregon, where I was a junior at Reed College. After class the previous day, my handsome chemistry lab partner, John Hall, had asked if

I would like to join him and four other guys in an attempt on Adams. The previous spring, John had taken me on my first backpack trip. Ever since I had been eager to climb a mountain and spend more time with charismatic John, so I happily accepted his invitation.

We had begun our ascent at one in the morning so we could climb the hard snow slopes above timberline before the sun softened their surface. When I first put on my daypack and headed up, I began panting so loudly that John later confessed he had wondered if I would make it out of the parking lot. And now here I was, an out-of-shape nineteen-year-old girl from the flatlands wearing borrowed boots and pack, trudging up a mountain in the middle of the night.

At dawn, we stopped to get ready to go up the glacier. John helped me strap a pair of crampons—the metal spikes that keep a climber from slipping on ice—onto my boots. He showed me how to tie myself into the braided nylon climbing rope, explaining that it would catch me if I fell into a crevasse. I didn't know what a crevasse was, but John was so confident I didn't worry. John tied himself to the front of our rope, I attached myself to the center, and Mike took the end. Fred, Ron, and George tied themselves to the other rope. I liked the secure feeling of this umbilical cord connecting me to these strong, attractive guys.

We began moving up again as the first shafts of light hit the glacier and the hard white snow glittered as though sprinkled with tiny mirrors. Ahead of us was an icefield sliced by long, narrow chasms with walls of blue and green ice—crevasses! Veils of cloud hung suspended above green valleys far below. Carefully placing my boots in John's footprints, I practiced what he called the rest step: Step, breathe, relax. Step, breathe, relax. Slowly and steadily, my body adapted to the unaccustomed exertion and I began to feel peaceful and strong.

Then I felt a tug at my waist, heard the sound of vomiting, and looked behind me to see Mike doubled up over his ice axe.

"Mike usually starts throwing up when he gets this high," John said, walking over to me. "He needs to go down—want to go with him?"

"Down? Me? Why?" I asked. "I love it up here."

John told me about altitude sickness, explaining that some people like Mike have an elevation ceiling above which their bodies don't adapt.

"So what's that got to do with me?" I asked.

"Well, you don't want to push it," said John, beginning to look uncomfortable himself. "Your breathing didn't sound so good in the beginning."

"But now I feel great," I said. "This is the most beautiful place I've ever been."

"Well, actually, when I asked you to come with us, I thought that by ten thousand feet you'd have had enough and be ready to go down, too," said John. "That way, Mike would have company."

"You invited me to come thinking I couldn't make it to the top?" I tried to stamp my foot in outrage, but the crampon points stuck in the ice.

"Ten thousand feet's good for a first climb," John said. "I'm willing to go down, but the others haven't been here before. Do you want to keep going with three guys who don't know the route?"

"If it's that or turn back, I'll go with them," I said. I couldn't bear the thought of leaving this gorgeous place I'd just discovered.

"Okay, okay, I'll go down with Mike. I've climbed Adams lots of times," said John. "I hate to leave all of you up here, but I guess the route's easy enough."

I untied from John and Mike and attached myself to the other rope.

John gave us the chocolate bar he'd brought for the summit. Then he led Mike slowly back down the glacier. As he turned and waved, I felt a pang of regret to be going on without him.

While John had traversed the steep slopes slowly and rhythmically, my less experienced ropemates headed straight up. Before long I was desperate for air, but I forced myself to keep going. Several hours later we reached a rocky point that looked like the top, and I flopped down to rest.

"This is a false summit. The real one's just ahead," said Ron, who had become our de facto leader. "Let's keep moving."

I looked in the direction he was pointing and saw a higher peak far—very far—in the distance. "This is it for me," I said. "I'll wait here." With shaking hands, I untied myself from the rope and the others continued up. As soon as they were out of sight I unzipped my wool dress slacks—the only pants I owned, since, like most women in the 1960s, I usually wore skirts or dresses. I squatted and relieved myself with great satisfaction. I had been holding it for hours, too embarrassed to tell the guys I needed to stop.

Then I sank onto a comfortable rock and looked around. The other Cascade volcanoes rose above the valleys like towers above a medieval city. Turning on my side, I watched, fascinated, as an intrepid ladybug crawled toward me, its red and black wings dazzling against the dark basalt rock. It seemed extraordinary to find a small insect high on this icy ridge, and I realized it was equally astonishing that I was up here. Dozing in the sun, I was content, and not the least sorry that I'd stopped short of the top. Right here were the space and peace I'd craved since my claustrophobic childhood.

An hour later, Ron, Fred, and George returned, jubilant, from the summit. Standing up stiffly to congratulate them, I noticed the long shadows.

"If we're going to get down before dark," Ron said, "we've got to glissade."

"Glissade?" I asked.

"Just take off your crampons, sit on the snow, and slide," Ron said matter-of-factly. "Use your ice axe to steer, and if you start going too fast, roll over and push the pick of your axe into the snow." Ron sat down and gave a quick demonstration. Then he, Fred, and George slid out of sight.

I had no choice but to follow, my heart pounding. I stuffed my crampons into my daypack, sat down with my legs pointing down the slope, and jammed my ice axe into the snow next to me. As I eased it out, I began to slide. Within moments, I was careening downhill, out of control. Terrified, I rolled over on my stomach and thrust the pick of my axe into the slope. I slowed, but continued sliding. I heaved my whole body up over the top of the axe and pushed it down with all my weight. I stopped.

I grinned into the snow and waited to catch my breath before sitting up and continuing my glissade. Soon I discovered how to use the axe at my side as a brake and a rudder. Glissading was fun! Down, down, down I flew for thousands of feet. At first the rough, frozen snow felt cold and uncomfortable beneath me, but soon I didn't feel anything.

The light was fading when I reached the lower slopes. The snow was now encrusted with scree—small pieces of volcanic rock—and freezing hard. When I reached the end of the snow, the guys were nowhere in sight. I saw footsteps heading down the scree slope and, willing myself to stay calm, followed them to reach the trees just at dark.

"Over here," someone yelled from the forest. Relieved, I followed the voice to where the others were waiting in the pitch black—all the flashlight batteries were dead. I sat down on a big rock, more exhausted than I'd ever been in my life. After a short rest, I put my hand on the rock to push myself up. My hand felt wet.

Strange, I thought. A wet rock.

I rubbed my fingers together. They were coated with a thick, sticky liquid. I put my hand back on the rock. It was covered with the same substance. Then I put my hand on my behind—and felt raw, abraded flesh.

The pebbles and scree in the frozen snow had acted like sandpaper and worn away my thin wool trousers, my underpants, and finally, my skin. I hadn't felt a thing as the ice had anesthetized my bottom. I now understood why the others had leather patches sewn on the seats of their climbing pants.

Telling these guys I hardly knew that I had shredded my pants, not to mention my rear end, was unthinkable. I grabbed my black pettipants from my pack (for some unknown reason, I'd brought these sliplike shorts along) and pulled them over my tattered slacks. In the dark, no one noticed anything amiss.

"Let's go. We've got to keep moving," Ron yelled, heading into the dark woods. I lurched along behind him, too worn out to protest. Before long it was clear we were lost. I was so tired that all I wanted was to lie down on the ground and sleep. After hours of staggering through the nightmare forest, we heard a faint whistle. We shouted back and John Hall appeared, waving his flashlight.

It was three in the morning, more than twenty-four hours since we'd started. As we drove out the dirt road, I sat on the edge of the seat and moaned whenever the car went over a bump. I stumbled back to my dorm room and tried to sleep, but intense pain in my thawing rear end kept me awake. I woke my roommate, Nancy, and asked for help. A true friend, she started picking out countless small rocks.

"Your bottom looks like a gravel pile," she said. "You need to go to the infirmary."

While the doctor removed the pebbles from my flesh with tweezers, I told him about our adventure and how much I loved mountain climbing.

"It's going to be a while before you can climb again," he said. "You won't even be able to sit down for weeks."

As I lay on my stomach in the infirmary, I dreamed of glaciers and rocky peaks. I wrote my family back in Chicago a letter: "I just climbed almost to the top of Mt. Adams. It was the most beautiful place I've ever been and the best day of my life. The mountains are where I belong."

CHAPTER 2 ▲

Back of the Closet
Davenport, Iowa, 1950

Grandpa gives me a glass of lemon tea with honey—a special treat for a child in his native Russia. The doorbell rings. Grandma curses, runs into the kitchen, and comes out with a broom in her hand. She races to the front door and swings her broom wildly at the caller.

"Go to your room," Grandpa orders me. I move into the dining room and squat by a low window, where I see blue-suited legs and body dodging a flailing broom.

"After what you did to Gert, don't ever set foot here again, you worthless foreigner," Grandma screams. "Go back where you came from."

Her rage makes me feel sick to my stomach. Somehow I know the man on the porch is my father. Grasping the windowsill, I strain for a look at his face, but all I can see is his back as he turns away and hurries down the street. As Grandma continues to curse, he disappears. I wait for him to circle back. Surely he'll come back to me. But he doesn't. Finally I return to the living room and my untouched lemon tea.

"Grandma, why did you do that? Was that my daddy?"

"He's a sick man and he made your mom sick, too," she yells. "Don't talk to me about him."

"I want to meet my daddy," I say timidly.

"You don't want nothing to do with him. Before Gert married him, she was smart and strong and beautiful. Look at her now. Worthless."

I'm terrified to ask any more questions about my father. Who is he? Why can't he come see me? What did he do to Mommy? Why is he so bad? I feel guilty—for what, I don't know.

One sticky summer afternoon a few months later, I discover a stack of dusty photo albums in the back of the hall closet. The flies are droning and everyone is resting or away at work. I creep into the closet. Leaving the door

With my mother, Gertrude Blum, in 1946, when I was sixteen months old.

cracked open for light, I pick up the musty volume on the top of the stack.

Flipping through photos of Grandma at home, Grandpa in his grocery store, Mommy and her sisters when they were my age, I come to a picture with my name on it. My mother is wearing high heels with ankle straps and holds me in a blanket. Next come many more pictures of me. My face is round and smiling, my hair dark and curly. One catches my attention: I am sitting on my mother's knee, reaching out—but the left third of the picture has been cut off. I look for photos of my father, but I can find only my mother and me—on a bench, at an amusement park, on a porch. Always part of the picture is missing.

Hearing the banging of a screen door, I close the album and slip out of the closet, my heart thumping and my stomach aching. In our house, the only evidence of my father's existence is me.

A Man and a Mountain ▲

1964

I LEFT THE INFIRMARY eager to climb again. Although John Hall and I studied together, and were good friends, he didn't ask me to go to the mountains with him again. The other Reed climbers, almost gods to me, seemed unaware that I existed. For weeks after my slide down Mt. Adams, I had to sit on an inflatable toilet seat during classes; I guessed I didn't look like a mountaineer as I walked around campus with the toilet seat peeking out from beneath my navy cape.

When I asked John how I could learn to climb, he suggested that I enroll in the climbing class for PE. "And don't get me wrong," he said, "but you could get in better shape."

I blushed, assuming he was alluding to the twenty-five pounds I'd put on since starting college. After a childhood where food was hoarded and parceled out as a reward for good behavior, the unlimited, high-calorie cafeteria fare was irresistible. Free to eat what, when, and as much as I wanted for the first time in my life, I stuffed myself at most meals. Worst of all was my three-candy-bar habit. When I couldn't solve a physics problem, I was magnetically drawn to the candy machine outside the physics library, where I ritually bought and devoured large Mounds, Milky Way, and Three Musketeers bars in rapid succession.

I decided to take up running to lose weight and get fit. On my first attempt, I bounded out to the quarter-mile track, made it a third of the way around, and collapsed, doubled over with a sharp pain in my side. As I staggered back to my dorm, I realized I needed a plan.

John suggested I train with a partner and introduced me to Fred Rothchild, a cherubic, balding music teacher who was a good friend to Reed students. Uncle Fred, as we called him, was as out of shape at age fifty-one as I was at nineteen. We decided to jog every morning at six, and set a goal of running out and back along a mile-long flat street called Reed College Place. We began by walking one block and then running one block for the entire two miles. We repeated this every morning for a week, usually in the rain and the dark, Fred entertaining me all the while with heroic sagas of Reed climbers. The second week we ran two blocks and then walked one. Running one additional block each week, we were able to jog the entire two miles after two months. We rewarded ourselves with a large cheesecake from Rose's Delicatessen.

At the same time, I attended weekly classroom sessions of the Reed College climbing course. When I met the instructors at the first outdoor training at Horsethief Butte in the Columbia River Gorge, they smiled knowingly. Assuming they'd heard of my misadventure on Mt. Adams, I vowed to make a good impression.

One at a time, the other beginners and I made easy ascents up the rock face, a top rope protecting us from a fall.

My encouraging instructor, Lucille Borgen, asked if I wanted to be the first to try the Swing, a more difficult route.

"Sure," I said, delighted to be singled out.

Roping up and scrambling to the bottom of the steep face, I was aware of the other students and instructors eating lunch across from the wall—all watching me. As my stomach turned somersaults, I carefully put the toes of my boots on the small bits of protruding rock and, using my hands for balance, tried to inch my way across the face. I was protected by a top rope anchored above and off to the side. My hands began to sweat and my legs to tremble. My toe slipped. I grabbed frantically at the rock, trying to regain my balance. I slid off the face, flew through the air on the rope, and slammed hard, shoulder first, into the adjacent rock wall. I heard giggles as I swung back and forth like a giant pendulum. I laughed in embarrassment as I was lowered to the ground and climbed up to the starting point to try again. Once more I slipped, and once more I hit the wall with a loud thump. I shook all over and my legs felt like rubber. But I tried again. And again. The climb became more difficult as I tired, but I maintained my optimism, believing that if I tried just a little harder, I could succeed. Finally exhaustion forced me to quit. The next week, I returned with John Hall and, with just him watching, easily made it to the top.

My reputation was not enhanced by an attempt another beginner and I made on Mt. Hood on a stormy November night. I was so thrilled when he suggested the climb, I didn't bother to check the weather. Starting at one in the morning from Timberline Lodge in a blizzard, we tramped around for an hour, gave up, and just managed to follow our steps to our car. Returning the next morning for some skiing at Timberline, we saw traces of our wandering footsteps, which had circled the parking lot but gone no higher.

To the Reed climbers, I was the tall girl who slid down Mt. Adams, walked around with a plastic toilet seat, swung multiple times on the Swing, and couldn't find her way out of a parking lot.

So no one invited me to go climbing with them until one memorable clear, cold February night. As John walked me back to my dorm, he was silent and morose. I asked him if something was wrong.

"I'm so bummed. I can't stand it," he blurted out, his voice anguished.

"What's the matter? Can I help?" John had never complained to me before, or expressed much emotion of any sort.

"I think my parents are going to get divorced," he said. He went on to tell me that his mother, who'd had four sons in six years, was tired of being the perfect housewife and was threatening to leave his dad.

I was touched by John's opening up to me and by his vulnerability. Wanting to support him, I confided that my parents had gotten divorced when I was three and I hardly knew my father. I told John I'd never met another child who didn't have a regular family, and had always felt painfully different from other kids. John put a reassuring hand on my shoulder. When he touched me I moved away, alarmed by how much I wanted him to come even closer.

I found it difficult to imagine that John's proper Anglo-Saxon Protestant parents had problems like those of my eccentric, loud Jewish family. Ever since I'd had dinner at John's house the previous Thanksgiving, I'd envied him his ideal suburban childhood, right out of *Father Knows Best.*

"One thing I do know," John went on, surprising me further. "I'm not going to spend my life living in a perfect house in the suburbs, working at a boring nine-to-five job and watching TV every night."

I couldn't believe his rejection of what seemed to me the American dream.

"What do you want?"

"I want to go on expeditions to Alaska, Peru, the Himalayas. I want to do research on glaciers in Antarctica and fly a spaceship to the moon."

I was awed and a little unsettled. John's passion and ambition both attracted me and made him seem unattainable. How could someone who dreamed of going to the moon want to spend time with me?

Then John abruptly changed the subject, asking me the question I'd been hoping for. "Do you want to climb Mt. Hood tonight?"

Yes! I ran back to my room, put on my new climbing pants (complete with leather seat patch), and grabbed my shiny new ice axe and gear. We reached Timberline Lodge just after midnight and went inside to fill our drinking bottles with hot water so they wouldn't freeze. As we headed up, the stars and white snow gave us our only hint of light; the mountain looming in front of us blacked out the stars from a third of the sky. With slow, rhythmic steps we moved toward the massive peak. John's steady, even pace was easy to follow and I was much stronger than I'd been on Mt. Adams. We reached the icefield well before dawn.

"There may be crevasses ahead," said John, taking off his daypack. "It's too dangerous to go on in the dark. We'll bivouac here until dawn." John and I huddled together in the lee of a huge rock, the wind whipping the surface snow around us. As the first golden light of morning traveled from the top of Mt. Jefferson to the other Cascade summits and finally to us, I fell in love for the first time. I was in love with John, with Mt. Hood, with mountains, with life, with love itself.

A fierce gust of wind hit us. "Let's strap on our crampons and rope up," John said, giving me a brief hug. "It's time for the top."

As I stood, a blast of ice crystals stung my face. Heading upward, I leaned into the gale, resolved to make it to the summit no matter what. Conscious of the rope around my waist connecting me to John, I followed him step by step up steep slopes festooned with delicate foot-long ice feathers. I was gasping for breath and my legs felt leaden, but I could see the summit getting closer. I took two breaths, lifted my foot, placed it in the step John had made in the ice slope, and relaxed. I took two more breaths, picked up the other foot, placed it in the next step, and relaxed.

As I repeated this sequence over and over, I thought about my family, and how afraid they were for me to swim in Lake Michigan, ride a horse, or even cross a busy road. And now, unbelievably, I was nearing the top of an 11,235-foot icy peak.

At ten in the morning of February 2, 1965, I stepped onto my first mountain summit. Around us, the symmetrical snowy cones of Hood's sister volcanoes—St. Helens, Rainier, Adams, Jefferson, and the Three Sisters—floated above wisps of morning fog. Between them lay the dark green Cascade valleys, and far in the distance, the ocean. I wanted to sing, to dance, and most of all, to hug John, but I didn't dare.

John grinned widely as he handed me a slice of brown bread and Norwegian goat cheese (the kind that doesn't freeze). We ate under the warm summit sun and then lay back on our daypacks to watch the clouds drift across the sky. I wanted our time on top of Mt. Hood to last forever, but clouds gathered and we had to descend. Facing the steep icy slope and kicking in the front metal spikes of our crampons, a maneuver called front-pointing, we inched our way down. As the slope became less steep and the snow softened, we took off our crampons, turned to face out, and plunge-stepped down the mountain. Unable to contain my

John Hall on the summit of Mt. Hood in 1963. He broke trail through knee-deep untrodden snow to reach the top. *(Photo by George Cummings)*

happiness, I belted out, "Oh, what a beautiful morning! Oh, what a beautiful day!"

The roar of an approaching engine interrupted my song and a Sno-cat filled with sightseers appeared. "Oh, look—mountain climbers!" one of them called out. The tractor stopped and cameras clicked. I held up my ice axe in triumph. The world saw me as a mountain climber—and it felt great.

With Susan Deery and another Reed climber on the way to Mt. Shuksan, 1966.

CHAPTER 3 ▲

Out of the House
Chicago 1951 to 1962

My family lives in a small row house in Jeffrey Manor on Chicago's South Side. We moved here when I was six, just before I started first grade. Grandpa said we'd just stay until my mother found me a new father.

In our home at 2136 East 100th Street, my three parents—Grandma, Grandpa, and Mommy—each have their own territory.

Grandma, a squat, stocky woman with short-cropped steel-gray hair, marches through the house with her chin thrust forward and her slip hanging just below the hem of her print housedress. She maintains the sanctity of her kosher kitchen with savage authority. If I break one of her rules, her eyes bore into me like bullets. "Oy! You're a worthless child," she barks. "No good." Slightly hard of hearing and addicted to television, she sits in her floral stuffed chair chain-smoking and shouting at the full-volume TV set. Often I sit with her and watch To Tell the Truth, I Love Lucy, *and* You Bet Your Life *until my head feels dull and my eyes ache.*

Or I go outside to the front porch, where Grandpa sits on a metal lawn chair searching for bargains in the newspaper ads. Having worked sixteen hours a day for decades in his own store, he's glad to be retired and spending his days outside reading the newspaper or talking to the neighbors. I hear him complain a lot: "Mayor Daley's a thief," "My daughter has no sense," "My wife's a bulldozer." I try hard to be good so he won't find fault with me, too. A tall, usually kind man with clear blue eyes, Grandpa gently wakes me for school in the mornings and offers me a glass of grapefruit juice. Sitting in his comfortable lap, listening to stories of his childhood in Russia, I'm secure. I will get good grades, I will praise him, I will do everything he asks so I can keep his love.

Mommy and I share a small upstairs bedroom. A tall, attractive woman and former concert violinist, my mother works long hours at a series of jobs—kindergarten teacher, hospital admitting clerk, librarian. When she's home she sits on her bed smoking cigarettes, drinking coffee, playing classical music on the radio, and burying me in an avalanche of words: "You're the most wonderful child in the world... You're so smart, much brighter than I am... You decide what I should wear... You have much better taste than I do... I read in your diary that you hate school. Why?... Get your clothes off the floor. I work hard to support you and you do nothing."

My grandparents are in charge of Mommy and me and intent on our improvement. They nag me to clean my plate, get my nose out of my book, stay close to home. They admonish my mom to save money, make friends, find a husband. And they yell at one another about their laziness, stupidity, and bad judgment.

When I can no longer bear the smoke, the television, and the fighting, I run away to the vacant lot down the street, sink back into the long grass, and dream of my escape. I search the clouds for my father, who must be waiting for the perfect moment to rescue me. Or maybe I can drift away to a place of my own—peaceful, quiet, expansive, and far from my family.

Higher and Higher ▲

1965

SPRING CAME TO PORTLAND in its usual style: sunshine, rhododendrons, azaleas, and roses replaced the winter rain and cold. I sat next to John in all my classes, my mind wandering from molecular dynamics and classical mechanics to thoughts about John and me together—very close together—above the clouds.

And that wonderful spring, I did become a climber. John and I ascended Mt. Hood and Mt. St. Helens by ever more challenging routes, and reached the tops of the Three Sisters and Mt. Washington. With every climb, I learned more about finding a route through a crevassed glacier, how to read the weather, when to go on and when to turn back. Every summit was a miracle to me.

One April day, crossing a gully on the steep Wy'East route on Mt. Hood, we heard the sound of falling rocks. Looking up, I saw huge boul-

I reach the 9,127-foot summit of Washington's Mt. Shuksan in 1966.

ders thunder toward us. Somehow we weren't hit. It had all happened so fast that I wasn't afraid; nothing bad could happen to me when I was in the mountains with John.

He looked ashen when we reached the top. "We could have been killed," he said. "I'm not climbing that route again." His words surprised me; I had thought of him as fearless. Years later I would realize that the high mountains could be as dangerous as they were glorious. But for now the mountains were a benign home, a place where I felt happier and safer than I ever had on East 100th Street in Chicago.

Between climbs, I lay in my dorm room reading mountaineering books, daydreaming about John, and reliving our times together. We had first met just after I arrived at Reed during the infamous Columbus Day storm of 1962, the most powerful windstorm to strike Oregon in the twentieth century. I was in my dorm trying to concentrate on Herodotus, with the roar of the wind getting ever louder. My roommate came running in. "The library roof just blew off," she said. "Come help save the books."

I headed out into the storm, leaning into a wind that tore at my clothes, the rain driving into my face. Out of the darkness a young man appeared. Together we struggled against the gale to the library, where he loaded a huge stack of dripping volumes into my arms.

"Sure glad I'm not high on a mountain right now," he said. "I was supposed to have climbed Mt. Hood tonight."

"You climb mountains?" I was impressed.

"Whenever I can," he answered. "When I tried Hood last time, a freak storm dropped so much powder I had to sidestep up the mountain on my skis. So much blowing snow, I could barely see my feet." As we worked all night to save the books, John shared his mountain adventures with me. I was fascinated by the foreign world he described and mesmerized by his deep blue eyes. His gaze captured me so intensely his eyes became my world. When John began to speak to me, my universe expanded to include his words, me, and him.

As John and I spent more time together, I usually found his eyes full of kindness and interest; they looked deeply into me, and saw and appreciated who I truly was. But sometimes, for no obvious reason, his eyes became distant and unfathomable, hard and glittering with ice. In the two-and-a-half years I'd known him, I'd never managed to figure out what caused this change in John—and I'd never dared to ask.

So on one of my morning runs with Uncle Fred, I took a deep breath and asked him if he had any idea why John was so changeable.

"Well, you know what Winston Churchill said about Russia?" Fred quipped.

"I've no idea. What?"

"'It's a riddle wrapped in a mystery inside an enigma.' He could have been describing John." Fred chuckled.

I laughed weakly and tried again. "What do you think about John and me?"

Silence.

"You make a great team in the mountains," Fred said at last. "I've never seen anyone as happy as John when he's plowing through deep snow in a blizzard. No matter how tough the conditions or the weather, he carries on quite cheerfully. You two do seem alike that way."

"Yeah, we both like being outside, no matter what. But what about us?" I pressed. "You know, as a couple."

Longer silence.

"I'd think twice about getting involved with John," Fred finally said. "He's fiercely independent. He told me he's determined not to get tied down, ever. Besides, I think he'd be one of the most difficult people on earth to live with. Things have to be done his way. You can climb mountains with John. But I wouldn't count on him for much else if I were you."

A few weeks later, at the end of spring break, I sat down next to John at dinner. Unexpectedly, and with a warm smile, John invited me to go up to the Reed ski cabin with him for the night. I nodded, my face flushed.

Driving to the cabin that night, we said little and pretended not to notice the electricity between us. Although we were twenty years old, I was inexperienced and alternatively eager to get to the cabin and then terrified of what might happen there.

Arriving at the ski cabin, John and I found the door locked.

"I guess we'll have to go back," John said, turning toward the car. I stood there with both relief and disappointment flowing through me. When I didn't start down the steps, he took the key out of his pocket with a wink.

We went inside and looked at each other. My heart was beating as fast as when I was near the top of Mt. Hood. John put his arms around me and kissed me for a long time. I melted against him. He began to unbutton my blouse. I so wanted to get closer to John, but his sudden shift from friend to lover was too abrupt for me. Uncle Fred's warnings about John and my mother's lifelong cautions about men came pounding into my brain. A wave of fear swept over me.

"I can't." I pulled away. "I'm really sorry."

After a long silence, while his eyes turned to ice, John turned away and opened the refrigerator. Wordlessly, we proceeded to burn two batches of fried eggs and half a loaf of bread before heading back to school.

Back on campus, I regretted my reticence, but was too shy to talk about it with John. We continued to study and climb together that spring, but he treated me strictly as a friend. I kept hoping something romantic would develop between us again, but I wanted John to love me, not just be my lover.

Having no plan for a summer in the mountains, I was happy to be offered a summer research job at Argonne National Laboratory near Chicago, where my uncle Sy worked. But I delayed my return home for a wonderful climbing trip up Mt. Olympus with John. I arrived in Chicago slender, tan, and fit.

"So, why are you a week late?" my grandmother complained in her Yiddish accent.

"John and I made the most amazing climb," I said.

"No more crazy talk about mountains and goyim," she said. "Aren't there any nice Jewish boys at your school?"

"John's my lab partner and we have all our classes together and we get the best grades in the whole—"

"I'm proud you've done well in your studies," my grandfather interrupted. "Now it's time you should have a quiet summer safe at home."

"I'm planning to climb on weekends at Devil's Lake," I said.

"Your job is important, not climbing," said my mother. "You need to be able to support yourself. I expected a man to take care of me. And now I'm a divorcée."

I stopped trying to explain, realizing they were never going to understand or approve.

Within a few days, I was focused on learning about brilliant yellow compounds called picrates. It was a relief to redirect my thoughts from my confusing relationship with John to straightforward scientific research. Not wanting to take time off, I got no exercise beyond expeditions to the Argonne candy machine and two weekends of rock climbing. My experiments showed some unexpected properties of picrates, and my advisor suggested that I write up my results for publication. I was thrilled. To allow more time to finish my paper, I canceled an ice-climbing course on Washington's Mt. Rainier I'd planned for the last week of the summer. John, who had recommended the Rainier course, wrote from Maine, where he was attending Outward Bound, "You are a FINK of the highest degree. I have no doubt you could have both finished your paper and gone to the Rainier school."

I appreciated John's confidence in me, although I didn't share it.

"If I never said it in a serious manner, let me tell you now," he continued, writing words I treasured. "You have the making of a very good mountaineer. You have both the physical ability and mental desire to reap the full rewards to be found in the highest mountains. Most important, you respond with love and empathy to the heights, not only as a gymnastic playground, but as a place of awe and splendor. And I've not encountered anyone else who has the enthusiasm and sense of excitement which I most admire about you."

My heart sang at John's words as I worked long hours in my lab. I completed my article for *The Journal of Chemical Physics* and returned to Reed for my senior year, out of shape once again.

Back at school, life was lonely. After just three years at Reed, John had left to attend Harvard Medical School, planning to get both an MD and a PhD. His parting gift to me was a perfect topic for my Reed senior thesis: collecting and analyzing volcanic gases.

An abrupt change in the composition of gases given off by a dormant volcano can signal an imminent eruption. In 1936, a Reed student had analyzed the contents of the toxic gases emanating from the fumaroles, or steam vents, near the top of Mt. Hood. In his thesis, he'd suggested that the fumaroles be monitored every fifteen years, and in 1951, another chemistry major had continued the research. It was now 1966, right on time for me to write a thesis combining my two passions, climbing and chemistry.

Every weekend I climbed Mt. Hood to collect fumarole gases from the summit crater, and during the week I analyzed their composition in the laboratory. I often climbed with Dr. Fred Ayres, a chemistry professor who was my thesis advisor and mountaineering mentor. A tall, shy, and very kind man, Fred had pioneered new routes in the Tetons, the Canadian Rockies, and the Andes. Fred told me about climbing in the Peruvian Andes with Claude Kogan, a superb French woman climber, and I began to dream of going on a big expedition myself someday.

John, meanwhile, was also lonely at Harvard. He wrote me a postcard each day, sharing his climbing ambitions and dreams of an

Collecting volcanic gases on Mt. Hood, 1966. *(Photo by Fred Ayres)*

extraordinary life. At Thanksgiving he called to say he missed the high mountains and me, too. He suggested we meet for winter break to climb three volcanoes near Mexico City—Orizaba, Popocatepetl, and Iztaccihuatl—the third-, fifth-, and seventh-highest peaks in North America. I'd originally proposed this trip to collect and analyze gases from the crater of Popo for my thesis. I was overjoyed, except now I had to tell my family I wouldn't be home for the December holidays. "I'm sorry," I told my mother on the phone. "I need to collect gases from volcanoes in Mexico."

"Mexico? Gases? Volcanoes? Is it safe? Who are you going with?"

"Two of my Reed climbing friends, John Hall and John Davies," I said. "We'll be fine."

"Arlene's going to Mexico with two boys," my mother relayed to my grandmother, who grabbed the phone.

"Two boys? Are they Jewish?" Grandma asked. "What Jewish boy will marry you if you run off to Mexico with goyim?"

Then my grandfather got on the line. "For this michegas we raised you?"

"It's not crazy," I said. "I'm twenty years old. I've been away from home for three years now, and I need to do this research for my thesis."

I could have been talking to three brick walls.

Just before Christmas, John Davies and I flew to Mexico City. John Hall met us at the airport and welcomed me with a long, warm hug. After months of getting his daily postcards, letters, and phone calls, it was hard to believe I was finally with him.

John showed John D. to their shared room at the inexpensive hotel he'd found for us. Then he helped me carry my bags to my room. When he unlocked the door and sat on the bed, talking about his first months at Harvard, my heart was pounding.

"You should come to Boston for grad school," John continued in his low, husky voice. "I can order us theater and symphony tickets for the season. And I'll show you the White Mountains. They're not so high, but plenty rugged."

"You really want me to come?" I could barely speak.

"I do," he said. "But it scares me, too."

"Why?" I asked, surprised that John, who was so fearless in the mountains, could be scared by me.

"I care for you a lot. If we're together all the time, I'm afraid we'll become so close, I'll no longer be a free agent. I'm determined not to get married while I'm a student."

We stopped talking. John's words and his nearness made me dizzy and short of breath. After a long silence John turned slowly and looked at me warmly, put his arms around me, and kissed me softly. This time it felt inevitable and right. Then he kissed me with more urgency. I followed as he took me to a place I'd never been before, trusting John's leadership in this Mexico City hotel room just as I did in the high mountains.

Awakening early the next morning, I felt calm and happy. Birds were singing outside the window and John was beside me, peacefully asleep.

I snuggled closer to him and he woke up and got out of bed abruptly. "We shouldn't do that again," he said as he quickly got dressed.

"We shouldn't?" I asked, a knot gathering in my stomach. "Why not?"

"I'm here to spend time with you and John Davies both," he said. "I'm not going to just shack up with you and leave John D. a third wheel to a couple."

I nodded, struggling to hold back my tears.

On a cool, clear day a week later, John Hall and I stood on the rim of the summit crater of Popocatepetl, at 17,887 feet an altitude record for me. From the top, we saw other snowcapped volcanoes towering above the brown Mexican plains. The area smelled of rotten eggs from the hydrogen sulfide in the fumarole gases seeping up from deep within the volcano. To collect these gases, I had six glass vacuum bottles that I had carefully transported on crowded Mexican buses to the mountain. We got to work, dangling rubber tubing down steaming fissures into the rock. When I opened the stopcock, gases from deep within the earth rushed inside the evacuated glass bottle. The chemical composition of these gases would enable me to predict whether this dormant volcano might soon erupt again.

When John and I said good-bye at the airport, he invited me to visit him in Boston over spring break to help make my decision about graduate school. "Harvard and MIT are great places to study chemistry," he said, giving me a quick hug. Although he'd shared a room with John Davies after that first night, my hopes for our relationship rebounded.

Back in Portland for my last semester at Reed, I analyzed the gases we'd collected. Unexpectedly, the gas composition suggested that both the Pacific Northwest and the Mexican volcanoes could erupt with devastating violence in the near future. After I finished my thesis, I wrote an article for the local Portland newspaper about a possible imminent eruption of one of the Cascade volcanoes.*

At the same time, I was deciding where I wanted to go to graduate school. Despite my family's opinion that I'd already had plenty of education for a girl, I knew I wanted to try for a doctorate in chemistry at a top school. I applied to Harvard, MIT, Caltech, and UC Berkeley.

---

*Fourteen years later, in 1980, when St. Helens blew its top, Marsh Cronyn, a professor on my thesis committee, told me he was impressed with the accuracy of my prediction. Popo, too, began to erupt again twenty-nine years later in December 1994.

In March the admissions letters arrived. I was thrilled to be accepted at all four schools and have the happy dilemma of choosing which one to attend. At Harvard, I would be near John and could join the legendary Harvard Mountaineering Club. Reading and rereading stories of its expeditions to Alaska and Peru, I fantasized about climbing with the HMC. Harvard it was, I decided, and went to visit Boston during my spring break.

After talking to a few chemistry grad students and professors, I looked up the HMC president. "My favorite books are about your expeditions," I told him. "You're the best. In fact, I'm coming to Harvard next fall so I can join the HMC."

"Well," he said, taking a step away from me and my unseemly enthusiasm. "Uh . . . actually . . . you're not eligible to join the club."

"I'm not?" I asked, unsure I'd heard him correctly.

"I'm sorry," he said. "We don't have women members."

"You're joking," I said. "This is 1966."

"I'm afraid it's true," he said, at least having the good grace to blush. "The mountaineering club is the last club at Harvard that's closed to women."

I stared at him, speechless. At Reed, being a man or a

With Dr. Jane Shell, my inspiring freshman chemistry professor and role model, in the lab at Reed College, 1963.

woman made no difference; everything was open to everyone. Why go to Harvard if I couldn't join the mountaineering club?

"Then I'll go to MIT," I told the HMC president. "I can join their outing club."

Now it was the HMC president's turn to look surprised.

When I returned to Reed, I wrote Harvard that I wasn't coming and sent my letter of acceptance to MIT. I'd been told MIT had never awarded a PhD in physical chemistry to a woman, so it would be a challenge. My brilliant chemistry professor at Reed, Jane Shell, had earned her PhD in organic chemistry there. I could follow her path.*

Because MIT's outing club did admit women, I reasoned, I'd be treated as an equal there. I couldn't have been more wrong.

---

*Jane Shell earned a PhD in chemistry from MIT in three years at age twenty-two and taught my freshman chemistry class at Reed in the spring of 1963. A creative scientist, eloquent teacher, and talented violinist, Jane was a role model for the four women students in our class. Inspired by her example, all four of us went on to obtain doctorates in chemistry.

CHAPTER 4 ▲

Girls Don't Read the Prayers

Chicago 1956

I go to religious school every Sunday morning with my neighborhood friends. We sit at old scratched wooden desks in a small bare room inside Congregation Kehilath Israel, the local synagogue. The boys wear yarmulkes, or prayer caps. I wear a felt skirt with a cat appliqué on the front and a sweater over my dickey collar. Because boys, but not girls, can go to Hebrew school during the week, the Sunday class is mostly girls.

Rabbi Mayefsky always calls on one of the three boys to read the prayers. One Sunday when I'm eleven, I listen to Stevie Gross's hesitant Hebrew with mounting impatience. I know I could do better.

That night I ask my grandfather to help me learn the prayers perfectly so I can read them in class.

"Well, the prayers are supposed to be read by men," he says.

"Please," I beg. "If you help me, I'll learn them so well that Rabbi Mayefsky will just have to call on me. Please. Please."

He nods his assent. With four daughters but no sons, Grandpa is delighted to teach me the Hebrew prayers he loves.

Every evening that week we sit at the red Formica kitchen table studying the Hebrew alphabet. Then for several weeks we practice the prayers. I love hearing my grandpa's melodious voice and imagine him in the yeshiva in Russia where he had studied Talmud. Once in America, Grandpa had no time for Hebrew. Since his retirement he has gone back to studying it and is delighted to have an apprentice. He is a dedicated teacher and I am an eager student. Finally he says I know the prayers as well as he can teach me.

The next Sunday, Grandpa walks me to my classroom.

"Don't worry about mistakes. You'll do fine." He kisses the top of my head.

I sit down, my heart pounding. Today is the day.

With my grandparents in front of our Chicago home on Rosh Hashanah, the Jewish New Year, 1954.

"Who would like to lead the Morning Prayer?" Rabbi Mayefsky asks.

I raise my hand. The rabbi picks Norman Kriloff.

"Who would like to chant the Shema?"

I raise my hand again. Grandpa had taught me to sing this prayer to a sweet melody. The rabbi calls on Nathan Ruben.

Each time the rabbi asks for a volunteer to read a prayer, I raise my hand. He always calls on one of the boys. I wave my hand and then my arm in the air, but I may as well be invisible. At the end of the class, I march up to Rabbi Mayefsky.

"Why don't you ever call on me?" I ask. "My grandpa taught me to chant the prayers perfectly. You always pick boys who don't know them as well as I do."

"Girls can recite the candle-lighting prayer," he says, "but the rest of the prayers are for boys." I stomp home, knowing the rabbi is wrong. Why shouldn't I read the prayers? God couldn't be that narrow-minded.

Grandpa is sitting on the front porch reading the Sunday paper, waiting for me. I tell him what happened.

"I know those prayers perfectly, better than those boys. It's not fair."

"You're right, it's not fair, but life isn't always fair," he says. "But it's not so bad. We can chant the prayers together here at home."

"I won't go to religious school if they won't let me say the prayers."

There is a long silence. "Okay," he says, his voice tinged with sadness. "You don't have to go anymore, if you don't want to."

"A Woman? Never!" ▲

1966

MY FIRST DAY AT MIT did not go well.

Leaving my dorm, where I was one of eight women living among five hundred men, I walked across Mass Ave. to the classroom building and headed down the very long main hallway. I began searching for my class among the labyrinth of corridors that led to scores of identical rooms labeled with multidigit numbers. I stood bewildered as serious young men surged past, books in their arms and slide rules in their pockets.

"Excuse me." I stopped a guy I recognized from orientation. "Do you know where quantum mechanics is?"

"You're taking 5.73?" He raised an eyebrow. "QM is a tough class . . . looking for a smart husband?"

I was speechless.

"Just joking. This way."

Swallowing my pride, I followed as he led me to the room.

After class, I needed to go to the bathroom but there was neither a women's room nor a woman in sight. Too shy to ask my fellow students, I wandered the bleak halls with increasing frustration until I finally spotted a secretary. Following her complicated directions, I eventually found the blessed ladies' room. Late for my next class, I entered the classroom as inconspicuously as possible, but four hundred pairs of male eyes turned toward me as I—the only woman in the engineering math lecture—made my way to a vacant seat.

That class over, it was time to visit possible research advisors. First on my list was a chemist known for his innovative work on molecular structure. In the grim basement of the same building I found the professor hard at work in the corner of his lab, surrounded by old bottles reeking of acid. I stood behind him for several minutes, shifting my weight and clearing my throat, but he didn't look up.

"Excuse me," I finally said in a small voice.

He frowned as he turned away from his dusty spectrometer. "What do you want?"

I smiled tentatively. "Um . . . I've come to talk with you about your research."

"You? Who are you?"

"Arlene Blum, from Reed College," I said. "I'm looking for an advisor."

"Forget it," he said, turning back to his work. "No girls in my group."

I felt like fleeing, but forced myself to stay. "Why not?"

"A waste of time," he said to the spectrometer. "We've never given a girl a PhD in physical chemistry. And we never will."

Holding back my tears, I left his lab and retreated to my dorm room.

Fortunately, my second day was more auspicious. I met Carl Garland, a professor who thought my gender no impediment to a PhD. Although his field, surface chemistry, wasn't my first choice, I was grateful for his support and joined his research group.

I also joined the MIT Outing Club, where, given the dearth of females at the school, women members were especially welcomed. With my new MITOC friends, I enjoyed rock climbing at the nearby Quincy quarries and rugged hikes in the White Mountains of New Hampshire.

John was occupied with medical school and had little time for mountains or for me. Occasionally, he would reappear in my life with an irresistible invitation. For the Thanksgiving vacation, he suggested we attempt a winter traverse of the summits of the White Mountains. I offered to bring the food.

My pack was huge as we set off in a snowstorm, but John didn't question its contents. After setting up camp near the top of Mt. Washington on Thursday night, I pulled a turkey, stuffing, cranberry sauce, and a pumpkin pie from my pack. John was flabbergasted, reward enough for my effort. Days of cold and blizzard contrasted with warm, loving nights in our tent. We had long, intense conversations about our lives and our dreams. But to my great disappointment, after this arduous adventure, John once again disappeared into his studies.

I missed him, but settled into a not unhappy routine of schoolwork, running, and MITOC trips. My classes at MIT were all that I'd hoped for, but the other students seemed to talk only about course work, exams, and grades. As the months went by, I increasingly missed the intellectual and political discussions I'd enjoyed at Reed. And as a woman studying chemical physics at MIT, I continued to feel myself a curiosity.

On the April day the Boston Marathon was run, I stopped by John's cluttered apartment to tell him the amazing story of Katherine Switzer. By registering as "K. V. Switzer," she became the first woman ever to receive a number and officially run this famous race. I'd heard on the radio that a monitor had spotted her and tried to pull off her number, but her shot-putter boyfriend had run interference. Katherine easily outran her pursuers, finishing near the front of the pack.

"They said running twenty-six miles would damage a woman's frail constitution," I said. "Can you believe it?"

"Boston *is* an awfully conservative place," John said. "You know I've never been happy here."

"Me neither," I agreed.

"There's something I need to tell you," he said, taking my hand. My stomach flip-flopped.

"Living here and going to medical school isn't right for me," he said. "I've decided to transfer to the Geochemistry Department at Caltech. I'd rather study rocks in the mountains than cadavers in the lab."

I was stunned. "What about me? What about us?"

"I'm sorry," he said. "My life has been awful lately. I had to make a change. Maybe we can do a big trip together next summer."

In a daze, I stumbled out of John's apartment and went to a market in Central Square to buy groceries. After a long wait to pay for my food, the clerk refused to take my out-of-state check. As I scoured my purse for enough cash, he yelled at me to get out of the line. What was I doing here, I asked myself, where the weather was so crummy and the people so rude?

To make my future in Cambridge even grimmer, the three other women chemistry graduate students in my class were all leaving. Each had different reasons, but the bottom line was that the MIT Chemistry Department wasn't a hospitable place for women students in 1967. I'd been considering transferring to the University of California at Berkeley earlier, but having missed the application deadline, I felt unhappily trapped at MIT.

That night I returned to my lab to finish an experiment. As I mixed the solutions, I felt depressed and alone in a sterile world of chemical reagents, burettes, and spectrometers. But I was not alone. Carl Garland, my sympathetic research advisor, was working late, too.

"Is something wrong?" he asked at the sight of my forlorn face. "Can I help?"

"No one can help," I said. "I don't belong here. I'm the only woman left in my class. I miss the mountains. I want to go to California, but it's too late to transfer to Berkeley."

He listened, thought a moment, and said, "Maybe I *can* help." He left the room.

Carl returned a few minutes later with a big grin. "I just talked to George Pimentel, my former advisor at Berkeley. He's a great guy, the chair of the Chemistry Department. He says you can come next year."

I couldn't believe it. How could Carl work this magic, and at eight o'clock at night?

"It's only five in the afternoon in Berkeley," he said. Carl went on to tell me that lots of women were getting PhDs there, the High Sierra was a great place to climb, and Berkeley had just been rated the best chemistry department in the country.

At first I was thrilled, and then wracked with uncertainty. Every time I heard "If you're going to San Francisco, be sure to wear a flower in your hair" on the radio, I wanted to go to Berkeley. On the other hand, doing something that people said that I couldn't do was so tempting—maybe I should stay at MIT and tough it out as the only woman in my class. Maybe I could show those professors they were wrong about women. But at what price? Was there much chance

I'd be happy in this alien world of gray suits, gray corridors, and gray days?

Berkeley, I thought, would be much more like Reed, where no one ever suggested it was unusual for girls to get PhDs or climb high peaks. I remembered my grandfather's support for my leaving religious school when I wasn't allowed to read the prayers. Compared with MIT, where proving something meant struggle and loneliness, Berkeley represented health and happiness.

I decided to finish out my year at MIT, put a flower in my hair, and go to California in the fall.

CHAPTER 5 ▲

Playing the Piano
Chicago 1953

School is so boring and my life so dull. Maybe I can learn to play Mom's violin. I pick it up and try to make music.

Grandma hears the noise and comes upstairs. "Put that thing back in the closet," she orders. "Come watch TV or go out and play like a normal kid."

I put the violin down with my heart thumping and go outside to Grandpa.

"I want to play the violin. Why can't I?"

"You know, your mother played so much she never learned how to live in this world. It wrecked her life. We're not going to let it ruin yours," he says.

"Sounds silly to me," I protest.

"No violin," he repeats.

My friend Carol Hagen lives in the next town house. After school a few days later, she pulls me into her living room to see the piano her family just bought. She'll have her first lesson that evening.

"Carol's taking piano lessons," I announce at dinner. "Can I learn, too?"

"That would be nice," my mother says, "but we don't have a piano."

I play my trump card. "Carol's mom says I can use theirs."

"No piano," Grandma says, spooning a large portion of lima beans next to the small piece of kosher steak on my plate.

I wolf down my steak and push the lima beans around on my plate to make it look like I've eaten some. "I'm going to play with Carol," I yell as I run out the door.

Carol is just back from her lesson with Mrs. Ellis over on Paxton Street and offers to teach me to play the piano. I'm thrilled. She shows me how to find middle C and to play "Twinkle, Twinkle, Little Star" on the upright piano in her dining room.

Then Carol wants to play Monopoly, but I want to practice. I play "Twinkle" over and over.

Carol's beautiful green-eyed mother welcomes me to practice at their house every day after school. The next week after her lesson, Carol again teaches me what she has learned and offers to teach me every week. I'm ecstatic.

I go to the school-supply store where I usually spend most of my weekly allowance of ten cents on penny candy. Feeling like I'm on a secret mission, I use my savings to buy a copy of a beginner's piano book.

Practicing every day, I work my way through the book. The weeks fly by, my pleasure at learning a new skill intensified by the delicious thrill of doing something forbidden. After a month I can play the first piano book and am ready to buy the second.

Carol's mother assumes my family will be delighted at my accomplishment. Imagining a happy surprise, she invites them over while I am practicing. My stomach lurches when all three of them file into the living room, but I play a rousing Jewish folk song and turn around with a big smile.

My mother, beaming with pride, claps and exclaims, "Wonderful! You did this all on your own! Such an amazing girl!"

My smile gets wider.

"Shut up, Gert," Grandma barks. "You don't know nothing. She's a sneak. We tell her no piano and what does she do? Behind our back." She grabs my piano book and throws it to the floor.

I start to cry.

"No more piano playing," Grandpa concurs more softly, putting his arm around me.

"Grandma and Grandpa love you and know what's best for you," my mother says, following her parents' lead. "You are a wonderful girl and you play so well. But you know enough piano now. No more."

"I don't want to play the dumb piano anyhow! I hate you all." I slam my

open hands down on the keys and run from Carol's house, the discordant noise ringing in my ears.

I climb up into the sour cherry tree in our backyard and stuff my mouth full of the tart, sweet fruit, spitting the pits down to the ground. I satisfy my hunger and keep eating because they taste so good, and then eat some more for comfort. After a long time, my stomach begins to ache and the branch on which I'm sitting pokes uncomfortably into my bottom. I hear Grandma calling me for dinner. I'm so close; it's surprising that they cannot see me. I hear my mom's music and the TV stop. It's getting cold up in the tree. I climb down and go into the warm kitchen for dinner.

Peru Adventure ▲

1967

"WOULD YOU AND JOHN like to climb with me in Peru this July?" I was thrilled to receive this invitation in a letter from my Reed advisor, Fred Ayres, during my last semester at MIT. I told John, who was just as excited. He came up with a plan to finance the trip with a grant for studying the fumarole gases from El Misti, a dormant volcano in the Andes of southern Peru. He offered to write the application to fund our 1967 expedition.

A few days later, he showed me his proposal. The expedition members listed were John, Fred Ayres, and Dick Birnie, a Dartmouth student with whom John had worked on the Juneau Icefield.

"What's going on?" I asked John. "You're not naming me as a team member?"

"Don't worry," he assured me, while avoiding my eyes. "A name on an application doesn't mean anything. I'm worried the committee won't take our proposal seriously with a woman on the team. They might think we're having a honeymoon rather than doing science."

"What are you talking about? The volcanic gas research was my thesis."

I couldn't believe John wasn't including me. He had always supported and encouraged my climbing and my science. Now he sounded just like that professor at MIT who didn't allow women students in his group. I was so hurt I considered staying home, but the opportunity to go on my first real expedition was irresistible.*

Flying into Lima alone, I watched the golden sun rise over the Andes. My spirits rose, too, for in a month I'd be climbing those shining mountains. Until then, while Fred Ayres, Dick Birnie, and John collected fumarole gases, I planned to explore Peru and Bolivia with Paul Pennington, a Reed climbing friend.

On our way to Huancayo, Peru, Paul and I changed trains in La Oroya, a copper-, lead-, and zinc-smelting town that was owned by an American company. With two hours between trains, we wandered through town, shocked by the disparity between the Americans' gated ranch-style homes and the grim cement-block hovels that housed the mine workers. Kids in rags begged us for candy and looked blank when I asked if they went to school. I had never observed such egregious inequity before, and was embarrassed at this face of my country.**

Paul and I spent the next weeks doing short climbs and treks around Huancayo, Ayacucho, and the Inca capital of Cuzco. Everywhere we went we were welcomed by the local indigenous people.

---

*John's assessment that he wouldn't have received funding if he had included me in the grant application is not so surprising given his environment. At that time, women eating with their colleagues at the Harvard Faculty Club were not permitted in the dining room; they were seated at a card table in the vestibule. Also, females were not allowed in the stacks at Harvard's Widener Library. I recently learned that John's volcanic-gas research grant was funded by the Harvard Travelers Club, whose members might well have had attitudes about women similar to those of the Harvard Mountaineering Club.

**In 2002, the Peruvian Ministry of Health reported that the average level of lead in the blood of children in La Oroya, Peru, was more than triple the allowable limit established by the World Health Organization. Nearly all the children in La Oroya suffered from lead poisoning, and almost 20 percent had blood-lead levels that required hospitalization.

Missouri-based Doe Run Corporation, the company that currently owns the smelter, agreed to modernize the plants and bring emissions down to acceptable levels by 2007. However, in 2004, the company requested and was granted by the Peruvian government a four-year extension for the implementation of its environmental-management plan.

Curious about Americans who would choose to walk through their villages, they generously shared with us the little food and drink they had. It didn't take long for us to understand the Indians' place as a conquered people in their previous empire. They were treated with scorn, not allowed to vote, and their land was stolen. Repeatedly we were told stories of ruthless hacienda owners who didn't hesitate to kill Indians who spoke up for their rights.

Once in Bolivia we began to hear about Ernesto "Che" Guevara, an Argentinean doctor working to educate and unite the Indians against their oppressors. Learning that Che was speaking at a political rally, I took a torturous all-day bus ride down into the deep valley of Los Yungas east of La Paz to see this legendary revolutionary. Sadly enough, Che didn't appear; looking at the heavily armed police among the crowd, I could imagine why.

Che Guevara was shot by the Bolivian military three months later. When I heard of his death, my heart was heavy. After seeing extreme poverty and injustice for the first time in the Andes, I resolved to do what I could during my life to make the world a more equitable place.

Finally it was time to meet up with John. Along with Dick Birnie, we trekked toward El Misti, the volcano where they'd collected gases. Then we headed to the Quebrada Llanganuca, a canyon northeast of Lima, and joined the rest of our climbing team: Bill Ross, a Stanford medical student who had gone to Reed, and Peter Schindler, a Swiss scientist at Argonne.

Our first mountain was Yanapaccha Norte, a lovely ice peak, about 18,000 feet high. Peter, our most experienced ice climber, led most of the way, chopping steps up the steepest sections. Leaving at 3:00 A.M., we climbed 2,500 feet to reach the top by midmorning and had a leisurely second breakfast on the narrow ice summit.

On the descent, Peter adeptly skied down a section of the steep, icy slope on his boots in a perfect standing glissade. Bill, his rope partner, was less experienced on ice, but attempted to copy Peter's elegant example. He slipped, lost his balance, and plummeted straight down

Hiking toward El Misti volcano in Peru with John, 1967. *(Photo by Dick Birnie)*

toward Peter. As the rest of us watched in horror, Bill knocked Peter off his feet and the two of them tumbled out of sight down a seemingly vertical ice gully.

Somehow we managed to rapidly descend the 60- to 70-degree slope following the path of their fall. (For comparison, a staircase is typically about 25 degrees. At 60 degrees you can reach your arm out in front of you and touch the slope.) We found them nearly a thousand feet below, stopped on a rocky outcrop in the middle of the glacier. Unbelievably, Bill was unhurt, but Peter had two broken ankles.

We lowered Peter down the snow and he somehow managed to scoot across the moraine on his hands and bottom. Bill went out for help and brought in porters and a stretcher. For thirty-six exhausting hours, we took turns carrying the stretcher across the rough mountain terrain. Finally a farmer with a horse came in to take Peter to the road, where he could get a taxi to Lima. As I sadly watched Peter ride off, I thought about the narrow line between success and disaster in the high mountains. No one had ever been hurt on one of my trips before, and the way a small error could snowball into a potentially lethal catastrophe left me shaken.

Bill returned to Lima along with Peter, while Paul, John, and I went to climb Nevado Pisco. Named after the national alcoholic beverage of Peru by the French team who made the first ascent (whether they had imbibed a few bottles on their climb is uncertain), Pisco was

Our 1967 expedition to the Peruvian Andes. *(From left)* Peter Schindler, John Hall, Bill Ross, and Paul Pennington in front of Huascarán. I'm the photographer.

considered one of the easiest 19,000-foot peaks in the world.

A long trudge over a moraine and up an easy snow slope brought us to a pleasant high camp in the saddle below Pisco. Summit morning, I awoke woefully ill with nausea and diarrhea. But I wasn't going to miss my chance to climb the mountain, a new high for me. Saying nothing about how sick I felt, I dressed and roped up.

Wearing nylon overboots without crampons to keep our feet dry in the deep, sticky snow, we plodded up. After several hours we reached the crux of the climb: a three-foot-wide crevasse that cut all the way across the slope just below the summit. Paul, the lightest in our group, went first and I belayed, or safeguarded, his jump across the crevasse by wrapping the rope leading to him around my ice axe which was planted firmly in the snow. With some difficulty he ran through the treaclelike snow, jumped, and barely made it across.

Next it was my turn. I focused my waning energy, tried to run, and leapt across the crevasse. I landed with my chest on the opposite side, my axe in the snow, and my legs in the chasm. My axe began to slip. In terror, I kicked my legs and tried to squirm forward, but I slid inexorably back down the snow and into the gaping crevasse. The rope at my waist stopped me with a bruising jerk. There I was dangling fifteen feet down in a huge ice cavern, and no one on the outside could see or hear me.

The rope around my waist kept me from breathing, so I put my feet against the wall to relieve some of the pressure. My smooth-soled overboots just skidded off. I tried not to think about the fact that a climber hanging with all her weight on her waist can survive only a few minutes, because the extreme pressure on the diaphragm makes it impossible to breathe. Then Paul lowered a second rope with a loop. I passed it through my waist sling and around my foot. I stood on the loop, and—finally, gratefully—took some deep breaths. My precarious situation notwithstanding, I was dazzled by patches of sunlight shining on the blue and green walls of the cathedral of ice that surrounded me.

John and Paul anchored the rope so I wouldn't fall farther and then I had to find a way out of the crevasse. I heard John's muffled

John amid the seracs in the Chopiqualqui icefall, Peru.

voice yelling to me but couldn't make out his words. Fortunately, we'd practiced crevasse rescue in the Reed climbing class. As I'd learned, I stepped up on the upper loop of rope with all my weight and pulled my body up. Then John pulled the lower and unweighted loop a foot or so higher. I stepped up on that. Paul now pulled his rope, which was lower. Repeating this over and over, I moved ever higher toward where the climbing rope cut into the lip of the crevasse. When I got near the top I yelled to John to help me get past the soft lip, the toughest part of climbing out. With John pulling on the rope at critical junctures, I managed to squirm out over the lower lip, and lay gasping on the edge like a beached flounder.

I was too exhausted from getting out of the crevasse to try to jump over it again. John barely made it across, also landing with his legs in the crevasse. But he succeeded in kicking his way out. Then he and Paul continued to the summit, only eighty feet above, as I lay wistfully watching. By midafternoon we were all back at camp; I felt better, but a little sad that I hadn't been able to reach the top. For much of the summer I'd been feeling intermittently weak and nauseous and had noticed an unusual color to my excretions. I had said nothing to my teammates, hoping that if I ignored my symptoms they would go away.

When we returned to Base Camp, Bill Ross arrived from Lima with the news that Peter was recovering and had returned home. The four of us then found a route up the glacier of our third mountain, the beautiful 21,000-foot Chopiqualqui. With only a few days left before Bill had to fly home, we decided to carry double loads to save time. While John and Paul descended to Camp I to collect a last load, Bill and I were supposed to carry eighty-pound packs to Camp III, pitch the tents, and make dinner. The final stretch to Camp III, at nearly 20,000 feet, was torture. With Bill not well acclimatized and me feeling more exhausted every minute, we moved in slow motion. Getting safely over a large, open crevasse took more than an hour. After that, every step upward with my mammoth load was an act of will; I desperately wanted to sink down in the snow and sleep. I continued up, mostly because the energy necessary to explain my quitting was

John Hall carried an enormous load on 21,000-foot Chopiqualqui in Peru.

even greater than that required to take one more step. John and Paul caught up with us just before High Camp and were audibly annoyed that the tents weren't set up and dinner ready for them. Through a blur of pain and exhaustion, I heard John griping that girls in the mountains were too slow.

Trying to redeem myself, I cooked dinner and then, long into the night, sat outside, melting snow for water and feeling like a martyr. We were to start for the summit at dawn.

The next morning I was so weak I couldn't sit up. While John, Paul, and Bill went to the top, I dragged myself around camp, preparing a good meal for their return. I felt so wretched I didn't even care that I wasn't climbing. Something was very wrong.

When the others got back to High Camp and saw my state, my teammates, all medical students, decided I had parasites and needed to get medical treatment in Lima.

"Where is the doctor of parasitology?" I asked in hesitant Spanish at the hospital in Lima.

"Honey, you don't need a parasitologist!" a Peruvian doctor in a white lab coat exclaimed in excellent English. "You've got hepatitis."

"Hepatitis? No way!"

"Your eyes and your skin are bright yellow. It looks like a full-blown case to me."

"I've been traveling with three medical students," I said. "They'd have noticed if I had hepatitis."

The doctor laughed and led me off for my lab tests. The results came back: I was several months into a severe case of hepatitis. The doctor asked if I had dark brown urine and white feces. I nodded. I had been too modest to mention my peculiar excretions to the guys and had assumed they were caused by changes in my diet. Now I understood why I had not been able to carry my load as easily as the others. At least it wasn't because I was a girl.

The doctor told me that the only treatment for hepatitis was rest and diet. I needed to eat lots of carbohydrates, like bread and honey, but no fats or proteins.

The next day I told the doctor I'd consumed two loaves of bread and two jars of honey. He looked startled and said I should be well enough to fly home.

I decided instead to fly to Pucallpa, by the Amazon River, to meet up with John. During several pleasurable days exploring the jungle, we never talked about the problems of the summer.

John finally apologized during our flight back to the United States. "I'm sorry. I didn't realize you had hepatitis," he said. "It's amazing that you kept climbing."

I thanked him, glad to be acknowledged.

"When you didn't carry your share, I thought you were acting just like a woman. Now I know—"

"What are you talking about?" I interrupted.

"Well, you have to admit that women aren't as strong as men in the mountains," he said.

Rage at being left out of the fumarole research and at all the comments about girls not doing their share swept over me. Seething, I stared out the window as our plane landed for a short layover in Bogotá, Colombia.

The plane stopped and I stood up, shouldered my carry-on bag, and turned to John. "I'm sick of you," I said. "I hope I never see you again as long as I live." I marched off the airplane, heart pounding but head high.

The plane took off and my grand gesture left me feeling empty. The next flight for Miami was in two days and I had twenty dollars in my pocket. I went to the tourist officer at the airport to ask about places to stay. When I told him my budget, with a wink he recommended La Casa Arnada—which turned out to be in the red-light district.

That evening I spent time with Luisa, a young employee at the hotel. It was comforting to talk with another woman. I confided to her how angry I was at John, and though I didn't think she understood much about research grants and climbing ice with a heavy pack, she did know about problems with difficult men. There was a knock on the door and Luisa said she had to go to work, leaving me to spend the rest of the night alone in a brothel in Bogotá.

The morning of my flight I got up very early to see the view from Monserrate Church on the valley rim above the city. Walking toward the edge of town at dawn, I asked a woman selling oranges how to get there. She looked alarmed, spoke to me in rapid Spanish, and took me to a police station.

The policeman, emphatic that I not walk to the church because of the thieves lurking on the trail, drove me to an aerial tram station and told me to wait until the trams started running in an hour.

After a few minutes, I became impatient. How could there be robbers so close to the capital? Why spend the little money I had for a tram when I could walk? I left the station and followed a wide trail leading up the hillside. As I ascended, the forest became denser and I felt uneasy. I saw several ragged-looking men in the distance. What if they were thieves like the policeman said? Off to my right a faint path led directly up a rocky face to the church. I ran over to the rock and headed up, glad to be out of the woods and away from possible bandits.

The rock face became steeper and the path more indistinct. I found myself doing climbing moves in a dress and smooth-soled shoes. As I reached for a hold my purse strap slid off my shoulder and my bag plummeted down the rock face into the thicket below. I looked down in disbelief. My camera, plane ticket, passport, and money were all gone. Bandits forgotten, I climbed down the face and started searching through the brambles. Nothing. After thrashing around for an hour, I stumbled back down the path to the city. My face was scratched and dirty and my dress torn.

A motherly Colombian woman caught sight of my disarray. "What happened?" she inquired with concern.

"I've lost my money, my passport, my camera," I said. "Everything."

"Poor señorita!" she exclaimed. "Attacked by the bandits in the forest."

"No," I tried to explain. "I dropped my purse in the bushes."

She was very sympathetic. "The bandits. This happens so often. Nothing can be done. You must let the doctor examine you."

I explained again that I had not been attacked and didn't need a doctor. I had lost my purse in the bushes. She smiled. "Then something can be done." She led me to a nearby school and spoke to the headmaster.

"The class will go back up with you," he said. "They will find your purse."

I, the tall American, felt like the Pied Piper as I led the thirty twelve-year-old boys to the thicket. They swarmed through the brush looking for my purse. Finally, a happy yell: *"Aquí, aquí!"* A boy with shining black eyes held up my purse and waved it in the air. The contents were intact. The Andean gods were with me.

"Muchas gracias," I said, giving the boy the few coins I had left. *"Muchas, muchas gracias."*

It was 11:30 A.M. when I got back to the hotel. My plane left at noon, and the next flight wasn't for another two days. Grabbing my things, I jumped into a taxi and made a dash for the airport. Fortunately, the plane was late and departure procedures were casual. I jogged through the terminal and up the stairs onto the plane. The door closed just behind me. I stumbled down the aisle, my carry-on bag swinging wildly. Sinking with a sigh into my seat, I noticed everyone looking at my torn dress and dirty, sweaty face. But I didn't mind their stares. Despite crevasses, hepatitis, and a difficult boyfriend, I had enjoyed my first real expedition. I had climbed above 20,000 feet. I was ready for Berkeley, California.

CHAPTER 6 ▲

The Lab School

Chicago 1954

Miss Benson, my fourth-grade teacher, asks me to stay after school. I'm worried. When we got a new reader this morning, I finished it right away and then went back to Dr. Doolittle, *lying in my lap. Miss Benson called on me to read aloud and I had no idea of the place.*

The forty other students file out of our portable classroom and Miss Benson hands me a sponge and a bowl of water. "You have to pay attention if you want to learn anything," she says as we both wash chalk off the blackboards.

I decide to tell the truth. "I try to listen, but class is pretty boring. Dr. Doolittle *is lots more interesting."*

"You're actually a very good student. Do you like school?"

"Not especially. I'd rather read books."

"There's another school you might like better. Have you ever heard of the Laboratory School at the University of Chicago?"

I say no, and she tells me that it's a nearby private school with smart kids and fun classes. Then she offers to talk with my family about my getting a scholarship to go there.

As I walk home, I decide I really want to go to the Lab School. I'll ask my parents right away.

I start with my mother, who usually follows my lead. "Mom, Miss Benson says I can go to a school with lots more interesting work. It's called the Lab School."

"I don't think so, Arlene," she says with a sigh. "The students there come from rich families with two parents. We're different."

"So, what does that matter? I want to go to the Lab School. Please," I plead, holding my folded hands up to her.

"Those children would go away on fancy vacations. It would make you too unhappy to see what the other kids have."

"I don't care about vacations," I insist. "I want to go and Miss Benson thinks I should."

"Miss Benson doesn't know you like I do. You'll be happier here in Jeffrey Manor with your friends."

Surprised that for once I can't boss my mother around, I go downstairs to Grandma. "What do you need with that smart-aleck school?" Her eyes remain glued to As the World Turns. *"Watch this; I just know he's going to leave her for his secretary. How many Jews go to the Lab School?"*

I go out on the porch to ask Grandpa.

"I'm sorry," he says, patting my shoulder. "That's too far for you to go every day. It's not safe."

"It only takes half an hour on the Jeffrey bus," I say.

"You're nine years old. That's too young to take the bus by yourself" is Grandpa's verdict.

Miss Benson does talk to my family the next day, but all three of them are in agreement for once. No Lab School.

The compromise is for me to skip fifth grade. The next fall I find myself, ten years old and bewildered, in the cruel social world of middle school.

Berkeley in the 1960s ▲

1967

ARRIVING IN DAZZLING BERKELEY, California, in the autumn of 1967, I felt as though I had landed on another planet. Students in brilliant tie-dyed clothing protested the Vietnam War. The sweet smell of marijuana, the acrid odor of tear gas, and the intoxicating voices of the Beatles and Bob Dylan filled the air. There could not have been a more dramatic change from the stark gray corridors of MIT.

Walking into physics class the first day of school, I was startled to see that the professor had hair down to his shoulders and looked about my age. He announced that the class was canceled. He was going to a demonstration at the Oakland Induction Center, and he urged us all to come, too.

An hour later I was in front of the Induction Center along with hundreds of others opposed to the war in Southeast Asia. Trucks filled with new recruits were arriving at the center, a first step on their path to Vietnam. "Hell no, we won't go!" we chanted as we linked arms and marched slowly in a large circle in front of the center. Joan Baez herself led us in "We Shall Overcome." Suddenly sirens drowned out our voices. Wearing full riot gear, Oakland policemen formed a line and advanced toward us as we continued circling and singing.

Each time I passed the phalanx of officers equipped with helmets, gas masks, riot sticks, and guns, I was terrified. Without warning, the police charged. I heard the sickening thud of a baton landing on the skull of a slight woman next to me and saw her sag to the ground. Much more frightened than I'd ever been on a mountain, I ran for my life. Dodging through unfamiliar streets, I found a bus heading for Berkeley and jumped on.

Back in my room at International House, I tuned my radio to a local station: "Joan Baez and ninety-six other demonstrators are being held in the Santa Rita jail on charges of disturbing the peace and obstructing the police." Assessing my first antiwar protest, I decided that although I admired the demonstrators' courage and shared their convictions, I was grateful to be safely in my room rather than in a jail cell. My first demonstration was my last for many years.

Finding a research advisor at Berkeley was also light-years from my experience at MIT. Told to check out a chemistry professor named Nacho, I visited his office on my second day. Entering a door labeled IGNACIO TINOCO, JR., PROFESSOR OF BIOPHYSICAL CHEMISTRY, I was greeted warmly by a wiry, energetic man.

"Blum," he said when I introduced myself. "You climb mountains, don't you?" He pointed to one of his walls, which was covered with dramatic photos he'd taken in the High Sierra. Then Nacho began to tell me stories about his adventures in science and in the mountains. His two passions were climbing and studying the structure of nucleic acids, the blueprints of life.

The four nucleic acid bases—adenine, guanine, cytosine, and thymine, commonly abbreviated as A, G, C, and T—combine to form DNA, which contains the genetic code for all living organisms. The arrangement of this simple four-letter alphabet on DNA helices specifies everything about us. Similarly, RNA is made up of the bases A, G, C, and uracil, or U. RNA molecules use the information from the DNA to make proteins.

Most DNA and RNA molecules contain thousands, or even millions, of bases in various combinations, and finding the order of the bases for a particular nucleic acid is a mammoth task. Nacho was studying the very small molecules called transfer RNA (tRNA), which contain a mere eighty bases. The ordering of the bases for several tRNA molecules from yeast had recently been determined. Nacho suggested that I try to predict the three-dimensional structure of tRNA, which would provide a key to understanding how these diminutive molecules worked to help build proteins in a cell.

At the end of our meeting, Nacho invited me to climb at Indian Rock with his research group. This steep crag in the Berkeley Hills served as a training ground for the climbers who pioneered vertical rock routes in Yosemite Valley. I decided then and there to work with Nacho. His interest in science and mountaineering attracted me, and his group had a warm, family feeling.

At the library that month, I pored over every paper I could find about tRNA. The order of the bases was known for ten kinds of yeast tRNA. I wanted to find a way to represent these sequences visually.

One day at Indian Rock, I noticed a climber wearing brightly colored Indian seed beads. I imagined the sequence of the bases of tRNA strung in beads. In a Telegraph Avenue head shop called Annapurna, among the hookahs and incense, I found what I was looking for: long strands of large red, purple, blue, and green seed beads.

Returning to the lab, I made a plan. Each G would be represented by a green bead and C would be blue; A would be red and U, purple. G bonds with C, and A bonds with U. I strung the bead sequence of a tRNA molecule on a piece of straight wire and bent it around so that blue beads were paired with green, and red beads were across from purple. My bead model folded neatly into the form of an H. With increasing excitement, I strung beads to represent the other nine known sequences of yeast tRNA. When I bent the wires to line up blue with green and red with purple, each time the shape was an H. The structure was simple and elegant. But was it true?

A 3-D model of my predicted structure for tRNA.

During my first year of graduate school, the base sequences for five more yeast tRNA molecules were discovered. I was excited to find that for each new sequence, the eighty bases folded into an H. Several scientists at Berkeley suggested that I write a journal article about my model.

Nacho was dubious, telling me that intuition was not enough; I needed experimental proof before publishing. He outlined some complicated experiments I could do to verify my model. I would need to purify some RNA molecules and break them into small pieces, whose base sequences should pair with the exposed bases in the tRNA. By trying to bond the small complementary pieces of RNA to the tRNA, I could learn which parts of the tRNA are exposed and which are protected and get clues about the molecule's structure.

It didn't sound like fun to me. After the quick gratification of constructing a model from beads, I was reluctant to turn to the tedious work of experimental science: measuring and purifying—and spilling and starting over.

Still, I had little choice but to do what Nacho advised. The best part of my research was buying big blocks of fragrant yeast from which the tRNA molecules were purified. I would walk through the pulsating streets of Berkeley to the Virginia Bakery, where I'd buy my yeast as well as large bags of éclairs and doughnuts. After sharing the treats with my labmates, I'd freeze the yeast cells, break them apart, and purify the tRNA molecules.

Our problems in Peru long forgotten, I'd often find myself daydreaming about John. Then I'd be rudely brought back to reality when my yeast solution boiled over. Still, I had other climbing friends with whom to share long weekends in the High Sierra. We'd drive to the mountains on Friday; hike to a Base Camp on Saturday; then climb, descend, and drive back to Berkeley on Sunday, arriving home in the early hours of Monday morning covered with mosquito bites or bruises. Monday and Tuesday I was exhausted. Wednesday and Thursday I was thinking about the next weekend climb. Friday we would be off again.

My second year at Berkeley, I moved into a house located on a scenic street called Panoramic Way, just above the campus. When I moved in, I had my choice of a large front bedroom with a breathtaking bay view for $60 a month or a small dark room in the back of the house for $40. I chose the latter to save my money for climbing trips.

Then I began studying seriously for my oral qualifying examinations. Many a promising graduate-student career ended during this dreaded ordeal. Mine were scheduled for the Wednesday after Thanks-

Traversing the outside of Lewis Hall at UC Berkeley with fellow chemistry graduate students Kay Wilkerson and Kathy Martin, 1969.

giving. I quit climbing to prepare for the exam that would determine my future. My housemates—three other women graduate students—understood what I was going through.

A week before Thanksgiving, I received a letter addressed to me in John Hall's familiar square handwriting. John and I hadn't seen each other since I'd stomped off the plane in Bogotá fourteen months ago. My hands shaking, I ripped open the letter. He wrote that he was happy at Caltech, but missing me. Would I like to spend the Thanksgiving holiday with him and his family in Portland and then ski on Mt. Hood? My heart raced. John still cared, and I realized I did, too.

I accepted John's invitation, knowing it wasn't wise to leave town the weekend before my orals. I rationalized that I could use the break, and I'd fly back early Sunday with a clear mind and have a last few days to review before the exam.

Thanksgiving dinner, skiing, and being close to John were all as warm and wonderful as I'd hoped. However, on Saturday, a heavy snowstorm blew in, closing the road and leaving us stranded on Mt. Hood. I was so happy floating down the mountain in the deep powder and nestling by the fire with John that, for the moment, I managed to forget my exams.

By the time the road opened Sunday afternoon, I'd missed my flight. Monday evening I finally made it back feeling anxious about my orals. I awoke Tuesday morning with an acute burning sensation and the need to go to the bathroom every few minutes. I was panic-stricken when the doctor diagnosed a urinary tract infection—the ill-timed reward of a romantic snowstorm. Orals usually last three or four hours, which in my condition would include multiple trips to the bathroom. How could I take my exams the next day?

There was nothing to do but confess my problem to the shy young chairman of my committee. He looked away with embarrassment and told me I could leave the room whenever necessary. An hour into the exam, I asked to be excused, ran to the bathroom, and used it in record time. Then I waited outside the closed door to the examination room. What was going on? What were they saying?

Why didn't they call me back in? I needed to go to the bathroom again but didn't dare leave.

The door finally opened. But instead of inviting me in, the examiners filed out with broad smiles and shook my hand, congratulating me on passing my orals. Was I imagining it or did the chairman of my committee wink when he took my hand?

The other students and faculty in the Chemistry Department were impressed when they heard I had passed in an hour—the shortest time anyone could remember. I resisted the urge to explain that my brief orals were probably as much about physical infirmity as mental agility.

Exams behind me, I was ready for a mountain adventure. Living in my little room and doing without a car, I'd managed to save enough money to pay for one. I was thrilled to hear that a biophysics professor at UC Berkeley was leading a trek through villages in the Annapurna region of Nepal. Having read every book I could find about Nepal, I longed to trek in this ancient Himalayan kingdom.

"We'd be happy to have you along," the leader responded when I inquired about joining his trip, "but there's no other woman for you to tent with. I'm sorry."

I couldn't go because of tenting arrangements? I was dumbfounded. Was there something wrong with me or with the world? I went for a long walk by myself in Strawberry Canyon above Berkeley and convinced myself I didn't care. I'd keep saving my money for another trip.

Several months later I heard about an expedition to Koh-i-Marchech, an unclimbed 21,000-foot peak in a remote valley in Afghanistan. A friend was the deputy leader. He suggested I apply to join the team, noting that I had more high-altitude experience than most of the other applicants.

My climbing partner Bill Dimpfl and I both applied. Bill, who had never ascended above 14,000 feet, was immediately invited to be part of the team, but I heard nothing.

Several months later I received an apologetic note from the leader. In spite of my excellent qualifications, they had decided that having a woman on the team might adversely affect the "camaraderie of the heights" and cause a problem in "excretory situations high on the open ice." I was being excluded only because I was a woman, and there was nothing I could do about it.

After indulging in a fit of tears, I decided I didn't want to go on their stupid trip anyhow. I hoped they'd have the worst climb ever. I hoped everyone got dysentery and were struck by lightning. I kept visualizing more and more disasters to befall the loathed expedition. They would starve. They would freeze. Avalanches would rain down on them.

The Koh-i-Marchech expedition was not successful. Many of the team came down with dysentery and my friend Les Wilson's tent was struck by lightning. He was so frightened that he jumped out of his sleeping bag and ran down the mountain in his stocking feet, pulling the tent behind him.

When Les told me these sorry tales on his return, I remembered with guilt my vengeful thoughts at being excluded and resolved never to wish harm on others again.

Continuing to look for an expedition that would welcome me, I called a local adventure-travel company for its brochure for a guided climb up Mt. McKinley, at 20,320 feet, the highest peak in North America.

Called Denali, "the Great One," by the local Native Americans, Mt. McKinley is the only mountain higher than 20,000 feet in a polar region. I had read that the spinning of the earth makes the atmosphere thinner close to the poles, so that conditions on Denali are equivalent to those on Himalayan giants thousands of feet higher. The mountain was much harder than anything I'd ever tried, but I was eager for a big climb. I was imagining myself high on Denali's arctic slopes when the brochure arrived, stating:

> Women are invited to join the party at base and advanced base to assist in the cooking chores. Special rates are available. They will not be admitted on the climb, however.

I read the brochure again, crumpled it into a ball, and pitched it into the wastebasket. Three pieces of fudge later, my dream of climbing Denali still vivid, I fished the brochure from the trash, smoothed out the wrinkles, and dialed the phone number at the bottom of the page.

"I'd like to join your Denali expedition," I began when the trip leader came on the line. "To try for the summit." I stopped and held my breath.

"I'm sorry," he said. "No women past Base Camp."

"Why not?" I asked, thinking he didn't sound one bit sorry.

"Women are not strong enough to carry heavy loads," he said. "And the high altitude. Women aren't emotionally stable enough to handle it."

I gripped the phone so tightly my hand cramped.

"I've climbed higher than Denali in Peru," I said, trying to keep my tone pleasant. "It was fine."

"You were climbing with a team of men?"

"Yes."

"The men did all the leading?"

"Mostly they led, but . . ."

"So the men did all the hard work?"

"They usually broke the trail through the deepest snow," I said. "But I carried a big load, too—and although I didn't know it at the time, I had hepatitis."

"I'm sorry, but women cannot climb Denali," he said, and then he hung up.

I put down the phone hard and ate the rest of the fudge. For two years I had been saving from my $212-per-month grad-student salary to go on an expedition. I wasn't going to squander my stash without a chance for the top.

I awoke at three the next morning with my mind whirring. *Did* the guys actually do more of the leading and hard work? I thought I'd contributed my fair share in Peru, but had I? When I was the only woman in a team of men, would people always question whether I was doing my share? I had to admit I wondered myself. Could women actually do a hard climb all on their own, without men?

An image of a team of strong women standing on the summit of Denali came into my mind. What if women made every decision, led the entire route, and carried all the loads? It would show that women are physically strong and emotionally stable enough to climb high mountains—or do anything else we wanted to do. Could I, a twenty-four-year-old graduate student who'd been climbing for five years, put together such a team?

At a local American Alpine Club meeting a few weeks later, I announced to a room of climbers, mostly guys, that I was looking for women to join me on an expedition to Denali.

"Chicks? Climb Denali? You must be joking," a man scoffed. "No way dames could ever make it up that bitch."

"Women don't have what it takes," a rock climber said. "Survive those killer arctic storms? No way."

"I'd love to go with you," a woman turned to me and said. I relaxed. "But I couldn't carry seventy pounds day after day," she continued. "We'd need men to carry the loads." I tensed again.

Discouraged, I relegated the women's Denali expedition to the proverbial back burner and decided to search for a more modest mountain adventure for the coming summer.

CHAPTER 7 ▲

A Path Through the Woods

Camp Pinewood, Michigan, 1956

My good friend Myra Lazerwith goes to Camp Pinewood every summer and I want to go, too. She tells me we can be cabin mates and talk all night. To my amazement, the summer I'm eleven Myra's parents persuade my family to let me go with her for a two-week session.

Camp Pinewood, a YMCA camp, sits on the shore of Echo Lake in central Michigan. I like making lanyards, canoeing, and roasting marshmallows around the campfire. But for a girl raised in an Orthodox Jewish home, the morning prayers in the chapel, the hymns about Jesus and the cross, and the bacon and pork at meals are most peculiar. When I push away a plate of creamed ham on toast, the girl sitting next to me asks me why I don't eat my lunch.

I explain. "I can't eat it because of my religion. It's not kosher."

"You can't eat ham?" She's incredulous. "I've never heard of such a thing. Are you Protestant or are you Catholic?"

"Neither, I'm Jewish."

"Does that mean you're a heathen?"

"What?"

"There are Christians and there are heathens," the girl says. "You're either one or the other."

"Well, I'm Jewish," I repeat. I don't know what a heathen is, but it sounds bad.

"I didn't think they allowed heathens at this camp," she says.

I slam my spoon down in the creamed ham so hard it splashes the tablecloth and a few droplets land on the girl's arm. I march out of the dining room and walk across the camp into a thick oak and maple forest. A narrow path takes me deep into the woods, dark and dank. Rays of bright sunlight filter through the trees. The cicadas hum and the birds sing. My feet crunch through the old, dry leaves on the trail. Has anyone ever been here before? Perhaps I'm following a fairy trail. When the path divides, I always take the right branch so I can find my way back. I am an explorer discovering new territory, strong, independent, daring.

The path ends in a clearing, a green meadow dotted with bright wildflowers. I sink down into the soft grass, feeling very far from Camp Pinewood, and at peace.

Real Women Climbers ▲

1969

JOGGING ON A FIRE TRAIL in the Berkeley Hills with my lawyer friend Frank Morgan in spring 1969, I saw a young woman jogging toward us in the golden evening light. I had never met another woman out running before and neither had she, so we stopped and exchanged phone numbers for a future run together.

"It's so great to see another woman out here," I told Frank. "Sometimes I feel pretty weird, liking to run, study chemistry, climb mountains." I told him about not being invited on the trips to Nepal, Afghanistan, and Denali, all because of my being a woman.

"You're fine," he reassured me. "Actually, I've been meaning to ask you on a climb. How'd you like to come to Canada with me and my old buddy Phil?"

"Wow! What peak? Where?"

"Mt. Waddington, the highest peak in British Columbia." Frank told me that he and Phil Trimble, an attorney friend of his in Washington, D.C., were joining a guided trip run by the Sierra Club and Mountain Travel to this 13,104-foot mountain, sometimes called Mystery Mountain. The team planned to fly into a small glacial lake in the wild Coast Range, hike through old-growth forests, cross huge glaciers, and try to climb Waddington's steep rock-and-ice summit pinnacle.

"I'd love to go." We finished our run at a sprint.

A few weeks later, Frank, Phil, and I landed in a small floatplane on an emerald green glacial lake surrounded by ancient rain forest. As we alighted from the plane, Bill, one of our guides, told us that this was the wrong lake and we'd have to bushwhack around a few ridges to meet up with the rest of the group.

"And worse luck, we have a woman with us," Bill said to the other guide so loudly everyone heard him.

I stood dead still as my face flushed bright red.

"What's the matter with that?" I asked quietly, feeling I had to say something—especially since everyone was looking at me.

"There are no real women climbers," Bill said.

"No real women climbers?" I repeated. "What's that supposed to mean?"

"It means that women either aren't good climbers or they aren't real women."

So upset that I couldn't even answer him, I turned away and headed up the trail.

We were to meet the other climbers and guides at Nabob Pass, which was "just around the corner," according to Bill. It was a hot, humid, and altogether miserable slog as we crashed through thickets of slippery slide alder and painful devil's club. Each time we rounded a ridge we saw yet another ravine full of thick brush. We'd gone around many corners, but Nabob Pass was nowhere in sight.

The swarming mosquitoes were so thick I had to keep my mouth closed to avoid swallowing them. I covered myself in foul-smelling repellent, which seemed to make me even more tasty to the mosquitoes. I stumbled, put my hand out for balance, and felt a sharp pain radiating up my arm.

"You shouldn't touch stinging nettle," Bill advised me. I struggled on, only to find more brush, more nettles, more mosquitoes, and no lake. As the light dimmed, we needed a place to camp, but there was no level ground. Finally, in the dark, we collapsed onto an open section of swampy trail that was free of forest but too lumpy to pitch a tent. Hungry and exhausted, I threw my sleeping bag onto the muddy, uneven ground, crawled inside, and prayed it wouldn't rain. As I fell asleep, I replayed Bill's comment about there being no real women climbers. I couldn't help wondering if he were a real male guide.

The next morning, after several more miserable hours of bashing through brush, we reached the legendary Nabob Pass, where the tents of our teammates were pitched beside an ice blue lake. The lake was surrounded by snow; a few small icebergs floated in the middle. Starving for breakfast, we were greeted by a thick cloud of mosquitoes even more voracious than we were. Hot, sweaty, and crazed by the buzzing, biting swarm, I dropped my pack outside the tent assigned to me, stripped to my underwear, and jumped into the freezing lake. For a few minutes I was relieved to be free of the maddening creatures but soon grew numb. I had to get out, and when I did the mosquitoes were waiting. I dove into my tent and met the other female on our trip.

Linda Crabtree, a twenty-one-year-old student from Los Angeles, was a nationally ranked fencer and a strong rock climber. A solid young woman with short sandy hair, Linda handed me a towel first and then some cheese and crackers. Just as she was telling me about how she'd ridden her bike up Vancouver Island carrying a full pack, ice axe, and crampons, Bill poked his head inside. "Is this the girls' tent? There's loads out here for them that's capable of carrying their share."

Linda laughed at his awkward attempt at humor, while I felt hurt. I resolved to try to stay as imperturbable as she was in the face of our guide's heavy-handed teasing.

From Nabob Pass, we bushwhacked down with huge loads through an almost impenetrable mass of tree roots, branches, shrubs, ferns, and devil's club to reach the edge of the Tiedemann Glacier. There, it seemed as though some giant child had thrown a tantrum, leaving in its wake an impossible rubble of ice, rock, and snow. Selecting a route through a jumble of blue and green ice blocks and towers was like finding my way through an oversize three-dimensional maze. I focused on the wild beauty of the ice formations, exactly where to put my cramponed boots, and my breathing. The tension of the last days melted away and I felt at peace.

For a day, we negotiated our way through a confusion of crevasses while carrying seventy-pound loads. We established our Base Camp on the Tiedemann Glacier across from Mt. Munday, named after Phyllis and Don Munday, an intrepid Canadian couple who pioneered a very long and difficult route from the ocean to Mt. Waddington in the 1930s and then climbed the mountain's slightly lower northwest summit.

Continuing above Base Camp, we found an intricate route through the Bravo Icefall, but were stopped short by a gigantic chasm between the glacier and the rock of the mountain. Called a bergschrund—or simply a schrund—this enormous crevasse is formed when the moving ice at the top of a glacier pulls away from the rock of the mountain, and also by temperature differentials between rock and ice. The schrund was too wide to jump across going up and extended so far in either direction there seemed no way to get around it. A frail-looking snow bridge crossed the gap. One of the guides moved cautiously across, belayed from below, while the rest of us anxiously watched. Then it was my turn. Taking a deep breath and clenching my ice axe, I stepped very carefully in his tracks, willing myself into weightlessness. I could hear snow falling from beneath my footsteps, far, far into the eerie depths below. I reached the far side safely. One by one the others stepped across the fragile bridge. As the last person pushed off his back foot, the bridge collapsed into the depths with an ominous thump. I felt a moment of fear and uncertainty about how we would get back down, but we were above the schrund now and could worry about our descent later.

Climbers finding a route through the crevasses and ice towers of the Tiedemann Glacier on Mt. Waddington, British Columbia, 1969.

After climbing for several hours along a steep ridge, our crampons scratching discordantly on the rock, we reached our first High Camp at Bravo Spearman Col. We dug a large snow cave and pitched two tents. The next evening during a wild storm, all twelve of us crammed into a Sierra Designs four-man tent to enjoy hot drinks and rousing songs. The following day we headed up and dug a smaller ice cave below the final rock tower. We spent the next day resting and acclimatizing, and then arose at 2:00 A.M. for our climb of the rock pinnacle.

After struggling into my frozen boots, I crawled out of the cave. In the dark, the dawn, and the golden light of morning we climbed ice-coated granite. The final chimney was extremely hard black ice, impervious to the points of crampons and ice axe. Summoning all my strength, I managed to fight my way up a steep crack and around two chockstones to emerge breathless at the top of the chimney. Easy rock scrambling led us to the summit of Mt. Waddington, the top of British Columbia, on an atypically gorgeous day. I was elated at reaching this perfect place after our long, hard struggle. Because of the dense forest, broken glacier, steep rock, and weather, Mt. Waddington is rarely ascended and is a prize for climbers. We had been lucky with the weather and could revel in the wild beauty of the surrounding peaks, glaciers, and forests. Rappelling down, I was acutely aware of the warm sun and the hard rock and my place in the world.

The summit pinnacle we climbed on Mt. Waddington, seen from the northwest summit we ascended the following day.

Back on the snow the *crunch, crunch* of our crampons was a resounding symphony.

Down at our ice cave, Bill congratulated Linda and me for making the summit in good style. We celebrated our accomplishment with a Jell-O cheesecake, the most delicious dessert I had ever eaten. Days later, we descended the glacier in a sitting glissade so fast that

Climber rappelling the bergschrund on Mt. Waddington.

we flew across the four-foot-wide open schrund without incident. Before heading home, Linda and I resolved to meet and climb together soon—just the two of us.

A month later, on Labor Day weekend, Linda and I spent a long weekend climbing sun-baked granite walls in Yosemite Valley. When we registered, the rangers were uncertain about the safety of a team of two girls, but by citing our Waddington ascent we managed to convince them. We began with the Overhang Bypass Route on Lower Cathedral Rock. Climbing as a rope of two women for the first time was scary but exhilarating. Before each pitch, we made a joint decision about the best route. I was eager to lead, as I'd not often had the opportunity. Linda, a better rock climber than I, had led before and let me go first this time. I found my way up the rock, the spray from nearby Bridalveil Falls providing cooling wafts of moisture. From the top of the climb, we looked down on the parking lot below, filled with people watching us through binoculars. As we scrambled back to the bottom we heard them exclaiming, "It's amazing! They're girls!"

Next, we attempted the more challenging Harry Daley route on a face called Monday Morning Slab. We were at the top of the climb when we heard crackles of electricity, our rope began to buzz, and the hair on my arms stood on end. An afternoon thunderstorm was on its way. As quickly as we could, we rappelled to a protected ledge near the bottom. A few minutes later, a couple joined us on the same ledge.

"Bill!" the woman exclaimed when she saw us. "Two women on the Harry Daley route! All alone!" They introduced themselves as Bill Isherwood and Dana Smith. As we huddled on the ledge waiting for the storm to blow itself out, they congratulated us on our climb.

I told them about trying to go on expeditions and my idea of doing a women's expedition, maybe to Denali. Dana said she'd like to come and Bill told us about two women from New Zealand who might want to join us. By the time the storm blew itself away, we had a plan and a nucleus for a women's team to attempt Denali.

I advertised for more climbers in *Summit* magazine and the Sierra Club's *Rock Climbing Newsletter*. And by Thanksgiving, we were six.

Linda decided not to come, but I had firm commitments from Margaret Young and Dana Smith, who lived near me in the San Francisco Bay Area; Faye Kerr and Margaret Clark, Bill's friends from New Zealand; and Grace Hoeman, a renowned Alaskan climber. I asked Grace to be our leader and when she accepted, she asked me to be the deputy leader. I was honored. We scheduled practice climbs for the winter and spring and solicited donations of food and equipment. The Denali expedition was becoming a reality.

Between weekends climbing at Yosemite and preparations for Denali, I worked hard on my PhD dissertation. A dozen more tRNA bases sequences—all of which fit the H model—had just been determined. I was invited to speak at a conference in Boston to present my model, but Nacho still wanted me to do more experiments before publishing my results.

Having not seen my family in Chicago for a year, I planned a visit during Christmas vacation with a stop for some skiing. I drove to Salt Lake City with my scuba-diving partner and good friend Howard Simon. We didn't arrive until the middle of the night, but undaunted, we bounded up to catch the first ski lift of the day. We skied for eight hours without stopping for lunch.

For the last run of the day, Howard suggested a rocky gully called the Corkscrew, which turned out to have barely enough snow to be passable. Exhausted, I followed Howard down. My ski hit a bare rock wall and I fell. I heard a crack and felt an excruciating pain above my right ankle. As the ski patrol took me down in a toboggan, all I could think about was Denali.

The doctor looked at my X-ray and announced that I had a spiral fracture of the tibia, the big bone in the shin, a nasty break. When I told him I had to climb Denali in June, he shook his head.

"There's not much blood supply there above the ankle and four months in a cast is average for this type of fracture," he said. "By then your muscles will be atrophied and your joints stiff. You'll need months of physical therapy. You won't be climbing any mountains next summer. And certainly not Denali."

It couldn't be true. My leg had to heal. I had to climb Denali. But the truth was my leg was so badly broken, I couldn't fly with the huge cast and had to cancel my trip to Chicago. Howard took the seat out of his VW Beetle so my extended leg would fit in his car and then drove me back to Berkeley. As my roommates were all gone for Christmas vacation, Howard moved into my house to help take care of me. We soon became more than friends, discovering that passion can coexist with a full-length leg cast.

Howard's straightforward friendship and devotion were welcome after my years of trying to decipher the complex John Hall. When we adopted two rollicking kittens and decided that we were a family, my life took on a measure of stability and love that I hadn't known before. And to my grandmother's enormous relief, Howard was Jewish and planning to become a doctor. I didn't mention he was working toward a PhD, not an MD.

The UC Berkeley infirmary found me an antique wooden wheelchair. For the next weeks, I sat in the chemistry lab in the old-fashioned wheelchair thinking hard—but not about chemistry. My intuition told me that the best way to heal my leg was to focus my mind on the broken bone. A picture of the area just above the ankle of my right leg was fixed firmly in my mind. For hours each day, I wiggled my toes and visualized the two jagged ends of bone coming together and mending quickly and well.

Exactly eight weeks later, I convinced my orthopedist to x-ray my leg to see how it was progressing. Reluctantly, he agreed. I held my breath as he came back into the examining room.

"It's incredible!" he exclaimed. "The bone is completely mended. I've never seen a spiral fracture heal so fast."

The cast could come off. My trip to Denali was on!

CHAPTER 8 ▲

Keeping the Peace

Chicago 1957

One Sunday when I'm twelve, my grandfather picks up a piece of chicken, takes a bite, and puts it down abruptly. "This is the toughest bird I ever tried to eat," he says. "After forty years, you still can't cook a chicken so anyone can eat it." He retreats to his chair out on the front porch.

The rest of us finish lunch. My grandmother mutters curses as she clears the table.

I follow her to the kitchen. "Try not to let Grandpa upset you," I say. "He didn't really mean what he said."

"That lazy no-good. Let him pluck and cook his own chicken!" she yells, turns the TV on loud, and plops down on the sofa.

"Shh. You don't want to upset Grandpa." And indeed, my grandfather, who was dozing on the front porch, wakes up and shouts, "Turn down that blasted TV!"

I run out to calm him. "So she wants me to pluck the chicken," he roars. "Then let her buy it herself. Whose hard work pays for every bite we eat?"

"Yours, Grandpa."

My grandfather throws down his newspaper. "And who's she to call me lazy?"

"She doesn't mean it. Her feelings are hurt."

"She's the lazy one. She never raises a finger except to turn up the volume on her damn TV. The dust of ages is under that couch she's lying on."

"Shh, Grandpa." I fear an explosion if Grandma comes out onto the porch. "No reason to yell."

"And tell your mother to turn down that music she's always playing!"

I rush up to our bedroom. "Mom, Grandpa's in kind of a grumpy mood and wonders if you might please turn down your music . . . just a little bit?"

"Turn it down? This is Beethoven's Ninth Symphony, the greatest music in the world. It's meant to be played loud. Just listen to this," she says, turning the volume higher.

"Maybe I'll close the windows so it won't bother Grandpa."

I pick up The Adventures of Sherlock Holmes *and lie on my bed, trying to ignore the loud music, the smoke, and the anger filling the air around me.*

The Damsels on Denali ▲

1970

"I'M FLYING YOU IN TO the mountain right away." Don Sheldon, our bush pilot, greeted us at the train station and began tossing our heavy duffels into his jeep. It was the summer solstice, June 21, 1970, and our train had just pulled into Talkeetna, Alaska, the base for flights into Denali. "Jump in and let's get going."

"I thought we were supposed to fly in tomorrow," I said. "Today we need to organize our gear."

"No time," he said, gunning the engine and pulling away from the station. "The weather's been terrible. I haven't been able to fly for a week. It's perfect now. Who wants to go on the first flight?"

"I do," Dana Smith Isherwood and I said at the same moment. Thirty-four-year-old Dana, a strong, outspoken chemistry teacher, had done a great job acquiring and packing much of our food and equipment. Just back from a honeymoon trek in Nepal, she wanted to get to the top of Denali and back down to her new husband as quickly as possible.

Dana and I climbed aboard, the ski plane took off, and we flew above a mosaic of forests, marshes, and meandering streams, the colossal white mass of Denali in the distance. Then the green water sparkling below us solidified into an enormous frozen river of ice, the mighty Kahiltna Glacier.

As we banked to the right around an immense icy buttress, the Denali Base Camp appeared below: a half dozen miniature orange

Talkeetna, Alaska, 1970, prior to our departure for the Kahiltna Glacier. *(From left)* Faye, Dana, Grace, Margaret Young, our bush pilot Don Sheldon, and me. *(Photo by Margaret Clark)*

and red tents and a huge peace symbol someone had stamped into the snow. Apparently, several other parties were there. With skis extended under the wheels, the plane bounced to a stop on a level area of packed snow at 7,200 feet, just outside the boundary of Denali National Park—where it could land legally.

A Japanese man with cracked lips and patches of skin peeling off his face ran up and embraced Don. For a week his team had been trapped at 14,000 feet. Three climbers were suffering from severe frostbite and needed immediate evacuation. We quickly hauled our gear from the plane as the injured men hobbled over, their hands wrapped in large white bandages. Smiling bravely, they bowed to us and climbed aboard. Don's plane taxied and vanished into the distance.

After exchanging a worried look, Dana and I proceeded to pitch our two orange nylon tents on the level snow, cook dinner, and organize our gear. As we worked and awaited Don's return with our teammates, Dana handed me a piece of bright red fabric.

I unfolded a flag she had sewn with our expedition name, DENALI DAMSELS 1970, in big white letters. I imagined six women wearing long skirts and carrying parasols, strolling amid the mighty glaciers of Denali—a more decorous image than the "West Buttress Broads," as some climbers had christened us.

By eleven that night, Don had made two more flights and all six of us—and our nine hundred pounds of gear—were safely at Base Camp. Trying to fall asleep in the Arctic twilight, I thought about the frostbitten Japanese and wondered what condition we would be in when Don flew us out.

Rising 18,000 feet from the surrounding tundra, Denali has been called the coldest mountain on earth. I knew that three years earlier, seven members of the Wilcox party had perished high on the mountain during a brutal nine-day storm. I pushed these thoughts away. Our team was well equipped and capable. If the weather gods were kind, I believed we could defy the dire predictions of the climbing establishment, reach the summit safely, and, very important to me, stay good friends. Finally, I slept.

The next morning dawned misty and hot. Light reflecting from the glacier was so intense that my eyes ached unless I wore both sunglasses and dark ski goggles over them. The best defense against altitude sickness is drinking liquids, so we spent many hours melting snow for water. In the background, unseen avalanches thundered across the glacier like a discordant symphony. No one mentioned the avalanches; there was nothing we could do about them.

After lunch, our leader, Grace Hoeman, a forty-eight-year-old anesthesiologist from Anchorage, called a strategy meeting. Grace had climbed more than one hundred peaks in Alaska, including more than twenty first ascents, many of them with her late husband, Vin Hoeman.

As we continued to melt snow, Grace outlined our plan. We would establish four camps on relatively level and avalanche-free sites along the Kahiltna Glacier and up the West Buttress. The Buttress, a steep ice and snow ridge, would provide us with access to the highest slopes on the mountain. Because our plane had to land outside the park boundary, our first carry went down 1,000 feet to the main Kahiltna Glacier and then back up to Camp I, four miles away at 7,800 feet. Camps II and III and High Camp would be at 9,700, 14,200, and 17,200 feet respectively. Each of us would ferry multiple loads of food, fuel, climbing gear, and personal equipment from camp to camp. We would need decreasing amounts of supplies at each higher level. With good luck and stable weather, we could reach the top in two weeks.

Early that evening, when the snow surface was frozen and easy to walk on with crampons, we decided to carry our first loads to Camp I. I tied several large boxes of food to my pack frame and tried to pick it up. It didn't budge. I gave it a few more tugs and then sat in the

Margaret Clark and I cook at Denali Base Camp.

snow, slipped my arms into the pack straps, and tried to stand up. As I leaned forward onto my knees, the bulky, top-heavy load pushed me facedown into the snow and pinned me like a squashed insect. I struggled free from the pack and the boxes fell off. Looking around, I saw my teammates grappling with their own loads and practicing similar peculiar calisthenics. I smiled and then giggled out loud. The giggle spread and became collective uproarious laughter.

Eventually, Margaret Young came up with a plan. Physicist, pilot, and airplane mechanic, thirty-seven-year-old Margaret was a brilliant and eccentric climber who did things her own distinctive way. At her suggestion, we paired up and sat down facing each other. Then we wrestled our way into our packs, joined hands firmly, and tried to pull ourselves to our feet without leaning too far forward. After several attempts, we managed to haul each other up. With intact seventy-pound loads on our back, we were victoriously vertical.

Margaret Young, Margaret Clark, and I tied ourselves fifty feet apart onto a climbing rope and set off down the glacier. Because Denali is notorious for sudden whiteouts that cut visibility to a few yards, we marked our route every hundred feet with a six-foot bamboo wand topped by a bright crimson flag.

My recently healed leg ached, and my mammoth pack pushed down on my shoulders and chest, making it hard to breathe. I shifted the load to my hips by tightening the waist belt of my pack. Then the belt cut painfully into the flesh above my hip bones. I loosened it and the weight pushed down on my shoulders and chest. Suffocating, I tightened the waist belt again. Feeling a little like an overburdened penguin, I managed to position the waist strap so it didn't hurt and I could still breathe, all the while standing very straight to keep the top-heavy load from throwing me off balance.

As I plodded along, a yank on my waist harness pulled me up short. I turned to see that Margaret Young, roped behind me, had stopped. As I struggled to keep my pack in balance while standing still, Margaret slowly peeled off her outer shell mitts, her inner woolen mitts, and her gloves, one at a time. She took her camera from a plastic bag and then from its case, held it up to take a picture, then

Denali photo © Bradford Washburn, Courtesy Panopticon Gallery, Waltham, Massachusetts

put it down and fiddled with the dials. Margaret eventually decided not to take the picture, repacked her camera with care, and put her gloves and mitts back on. We set off again, but before long I felt another tug indicating she'd stopped again. I watched with irritation verging on disbelief as she repeated the entire process with mitts, gloves, and camera.

Margaret Young would not be rushed, Margaret Clark had a pesky altitude cough, and I was a slow walker, so it took us five exhausting hours to cover the four miles to Camp I. By the time we dropped our loads and headed down, the sun had set for the short arctic night. Unburdened, our return trip was exquisite, the moonlight illuminating phantom peaks. We were silent, struck mute by the beauty around us. It was so quiet that even a whisper sounded like a roar. The cold had put a crust on the snow; as we stepped on it with our crampons, ice shards slid down the slope with a tinkling sound.

The next morning I awoke and joined what was to become a daily ritual in the crowded tent I shared with Grace and Dana.

"Where's my purple sock?"

"Who took my dark glasses?"

"Don't use that shovel to dig snow for drinking water. I used it for the loo."

"You're spilling your hot chocolate on my sleeping bag!"

Similar exclamations, punctuated by Margaret Clark's rasping cough, emanated from the other tent. But by the time we finished our oatmeal and cocoa, only Grace's dark glasses were still missing.

"I know I put my glasses in the tent pocket," Grace said, looking for long moments at each of us. "Someone has stolen them."

Paranoia about lost objects can arise easily in a crowded tent, so I wasn't especially concerned about Grace's accusation that we had a thief in our midst.

As the day became warmer, Grace took off her woolly red hat.

"Grace." I pointed, smiling, at the glasses perched on top of her head. The other women chuckled discreetly.

"It isn't funny," Grace muttered through clenched teeth. "If Vin were here, nobody would laugh at me."

Grace had lost her beloved husband the previous year in an avalanche on Dhaulagiri in Nepal, a tragedy that had also claimed Bill Ross, with whom we'd climbed in Peru.

"We were laughing with you, Grace," I said gently. Somewhat awkwardly, I squeezed her hand. "We all lose things up here. And I've been meaning to tell you how sorry I am about Vin."

"Thanks," Grace sighed. "Sometimes I think life isn't worth living without him."

Faye and I listened as Grace told us about her three-year marriage with Vin, the best years of her life. Then she seemed ready to be distracted from her grief.

"Who wants to climb the mountain wearing their long underwear?" I asked. This sunny morning seemed a good time to take the publicity pictures we'd promised sponsors who had donated food and equipment to our expedition.

Dana, Margaret Clark, and I took off our outer clothes and, wearing only our powder-blue insulated Duofold underwear, boots, and snowshoes, marched up the glacier.

"I dreamed I climbed Denali in my Duofold underwear," I joked, mimicking the old Maidenform bra ads.

"No more dreaming," Dana said briskly. "Let's start climbing." Fully dressed again, she hoisted her seventy-pound pack as though it were full of feather down. "This perfect weather may not last."

Grace had lashed two large boxes to her pack frame and tied snowshoes on top. She lifted the gigantic load, struggled to put it on, and suddenly dropped it back onto the snow. "It's too hot to climb," she said. "My head aches and I feel dizzy. Let's rest now and carry the loads tonight, when it's cooler."

"What? We're all set to go," Dana said, frowning.

"Dana, quit pushing!" Grace turned to face Dana. "We're not going so fast we get sick."

"If we're going to climb this mountain," Dana argued, "we need to take advantage of good-weather days."

"I'm leading this trip," Grace said, standing up straight and glaring at Dana. "And I say we're waiting till it's cooler. Don't you try to take over."

"I wouldn't if you were doing your job," Dana fired back. "You should be leading us up while the weather's good, not just hanging around complaining."

Grace looked stricken and then set her jaw. "I was warned that

Dana, Margaret Clark, and I model our powder-blue insulated Duofold underwear on the glacier.

you're impossible to deal with, Dana. You do what I say or you don't climb the mountain."

Dana spun away, stunned by Grace's words.

Fearing their anger and its effect on our expedition, I followed Dana back to her tent.

"Grace is mostly upset that she's feeling weak from the altitude," I said. "It's not you."

"Are you kidding?" Dana took off her pack and dropped it in the snow. "I don't trust her one bit."

"Come on. You know she's climbed dozens of tough mountains."

"Right. With her husband."

"Grace didn't mean it, Dana. She just lost her temper. Try not to take it personally."

Dana stood silently. I could feel her rage.

"You and Grace have to get along," I said quietly. "Otherwise we may as well give up and go home."

After a long silence, Dana shrugged her shoulders and shook her head. "Okay, okay. I'll try. But she better never talk to me like that again."

I hurried back to Grace, who was seething in our tent. "Dana doesn't realize how harsh she can sound," I explained.

"She's trying to take over."

"No, I don't think so. She's just in a hurry to get back to Bill."

Staring into the distance, Grace said, "That, I can understand. I miss Vin so much. At night I lie awake thinking if only I had been on Dhaulagiri I could have saved them. But they wouldn't let me come because I was a woman. Just like on Denali, two times they sent me back because they didn't want a woman to succeed."

I knew Grace had ended two previous Denali attempts at about 12,000 feet.

"I'm so sorry. It's dreadful," I said. "But we're here on Denali now and this time, no one is going to send you back because you're a woman."

Based on my own experiences of being excluded from trips because of my gender, I felt sympathy for Grace. And I accepted without question her story that she'd been sent back because of sexism. This was a possibly lethal error.

Grace and Dana apologized to each other at dinner and agreed to try to get along. Maintaining good relations among my companions was a job I readily took on after my early years of trying to maintain harmony among the three warring adults in my childhood home. To my relief, I was more successful at keeping the peace here on Denali than I had been on East 100th Street in Chicago.

During the next six days we ferried seven hundred pounds of food and equipment—sixty pounds at a time—to Camp II, at 9,700 feet, just below Kahiltna Pass and then to a cache 2,500 feet higher. The weather was good—for Denali. One minute the sun beat down fiercely. The next minute swirling snow whited everything out. I was usually wet from sweat, falling snow, or both.

Meanwhile, Margaret Clark offered to supervise the construction of a snow outhouse at Camp II. Clark, a gregarious geology student from Christchurch, New Zealand, was considered one of her country's best climbers. This dynamic thirty-five-year-old woman weighed just over one hundred pounds and could carry loads nearly two-thirds her body weight.

Following Margaret's lead, I stamped back and forth on the surface of a ten-foot square of snow to consolidate it. Then we cut the packed snow into two-foot-square blocks and piled them up to make a wall. Behind the four-foot wall, we dug two deep holes in the snow. Margaret Clark, wearing her signature floppy sky-blue sun hat and proper lady's white gloves, completed the construction by placing wands, their red flags flying gaily in the breeze, every foot or so along the top of the wall. No matter how intense the blizzard, we'd be able to find and use the loo in relative peace and privacy.

Nearby, Faye Kerr, who admitted to being about thirty-nine years old, was practicing yoga in the snow, wearing a sun halter and shorts. A native of Melbourne, Australia, Faye had extraordinary strength, patience, and tolerance to extremes of heat and cold. She was one of the first women to make climbing her life work. Blonde, beautiful, and easygoing, she had trained to be a teacher, but took any job that gave her the chance to climb. Earning money as a nanny, waitress, and hotel maid, Faye made dozens of demanding ascents in the New Zealand and European Alps, the Andes, and the Canadian Rockies, but she cared more about being in her beloved mountains than about reaching their summits.

As Faye did her yoga, one of the guys in another party camped across from us did sit-ups in the snow. I went over to meet our neighbors. A wiry man stepped forward and introduced himself as Mike Bialos, the leader of a team of six men from Seattle.

I smiled, told him who we were, and put out my hand.

Mike nodded and walked away. I put down my hand, wondering if I had said or done something wrong.

In the morning, we woke up at Camp II to find two feet of wet snow blanketing everything. Before we could pack for the long move to Camp III—nearly a vertical mile above us—we had to shovel the heavy snow off our tents and gear, an inauspicious beginning for our toughest day yet.

For hours, we picked our way through a chaos of icefalls and rock buttresses. Taking my turn at leading, I felt like a pioneer forging a path into unknown territory. Moving across the heavily crevassed glacier, I stayed as far as possible from the menacing hanging ice cliffs on the side. As we reached Windy Corner, a small campsite above the ice cliffs, the clear sky clouded over and thick, wet snowflakes began to float gently down.

Feeling fortunate to have gotten out of avalanche territory before the new snowfall, we sat down to eat and rest. Within moments, a thunderous roar sounded just below us. Ice blocks as big as houses broke off from the cliffs directly above the track we had just made and cascaded down the mountain. Wind and a spray of ice particles blasted us. Our footsteps from minutes before were buried by tons of debris. Waves of fear flooded through me. Breathe, I commanded my lungs. Breathe.

For long moments no one said a word. We understood that our survival depended as much on the random timing of avalanches as on our skill and hard work.

"This is a murderous spot," Margaret Clark said. "Let's move on."

We were still two thousand feet below camp. The wind picked up and snowflakes whirled around us like malicious dervishes. We had to keep going no matter how scared or tired we were.

I pushed myself to my feet and began moving upward. I counted my steps and when I reached twenty allowed myself a blessed minute to rest. Twenty steps, rest. Twenty steps, rest. Time moved slowly. Twenty steps, rest.

After hours of toil, the tents of another party at Camp III finally appeared in the distance. As I reached the top of the last snow slope, an apparition in a bright red parka handed me a steaming bowl of tea.

I pulled off my wet gloves, took the delicate bowl with both hands, and felt its delicious warmth on my freezing fingers. I took a sip. The tea was aromatic, sweet, and utterly delicious. The man in

Dana carries a seventy-pound load on Denali, with Mt. Foraker in the background.

the red parka, a member of a Japanese ski expedition already at the camp, smiled and bowed when I returned the bowl to him with my thanks.

Energized by the tea, I went to help unpack and pitch our two orange tents. As we worked, the snow stopped falling, the storm clouds cleared, and the sun began to set. The glorious vision of jagged ice mountains shining amid purple and pink clouds made me forget my exhaustion.

We moved into the tents and began the laborious process of melting snow and cooking. The dry, fluffy snow contained little moisture, so it took a long time to melt enough water for dinner. When we finished eating, Grace told us to turn off the stoves. "The noise and fumes are bothering me," she said. "I'm tired and I want to go to bed. We can finish melting snow in the morning."

"It's been a good day's work," I said, following Grace's lead even though I was still thirsty. "Once we bring up the loads from the cache, we'll have two weeks of food and fuel up here—enough for the rest of the climb."

"I could be on top and on my way back down to Bill in a week," Dana said with a grin.

"There's no hurry as far as I am concerned," Faye said, turning off the stove. "It's perfect right here."

The next morning I was awakened by snowflakes landing directly on my face. The moisture from our breathing during the night had condensed and frozen on the tent's ceiling, and when we touched the walls we triggered an indoor snowstorm.

I peeked outside and saw more snowflakes. It was too stormy for climbing, and I retreated into the soft warmth of my sleeping bag. After eight continuous days of carrying loads up the mountain, I was grateful for a rest. Today we would just eat, drink, and adapt to the altitude.

Grace passed around moose jerky for breakfast. When she told us proudly she'd made it herself, I asked her how.

"No big deal," Grace said. "After I shoot the moose, I carry it out, butcher it, and dry the meat."

I looked at her slim figure. "You can carry a dead moose out of the back country all alone?"

"No problem," she said. "I just cut it in quarters and carry them out in four separate trips. We used to hunt for moose and caribou every year together, Vin and I. Now I go on my own."

"No problem carrying a bloody moose quarter?"

Though impressed by her strength and tenacity, I thought hunting solo sounded risky. The fact that Grace had also done most of her recent climbs on her own also struck me. I thought back to the high-altitude physiology studies we'd taken part in at the University of Alaska just before the climb. On a simulated ascent to 20,000 feet in a pressurized altitude chamber replicating the atmosphere on top of Denali, most of us felt fine and could solve complex math problems. Grace, on the other hand, developed a migraine and couldn't do the math. I'd hoped she'd adapt better on the mountain, where she would gain altitude more gradually, but since we'd been high, she had been irritable and complained about headaches.

I'd noticed Grace didn't seem to be drinking much water. The three Americans on the team, Margaret Young, Dana, and I, each drank nearly a quart of cold water after downing the salty moose jerky. But Faye, Margaret Clark, and Grace were craving cups of hot tea. I had been in charge of planning the food and, naive American, had rationed one Lipton's tea bag per person per day. By the time we reached Camp III, it was apparent this was a serious miscalculation.

"No one can survive on one cup of tea a day," Grace said.

"I don't need to drink tea," I said, handing her my tea bag.

"And you can make two or three cups from one bag," Margaret Young suggested.

"More than one cup of tea from the same tea bag?" Even the gentle Faye shuddered at this philistine suggestion.

"That Japanese tea last night was first-class," Margaret Clark said. "So much better than that dreadful Lipton stuff. Arlene, maybe you could ask if they have any to spare."

I went over to the Japanese tent. "The tea you gave us yesterday was great," I said. "By chance do you have any extra?"

The Japanese climber nodded and with a wide grin offered me a big red and yellow box. Lipton's tea bags!

A neighboring climber shares his "authentic" Japanese tea.

Chapter 9 ▲

The Top of the Rock
Colorado 1957

July 20, 1957

Dear Mother,

Yesterday we drove from Iowa to the Badlands and Mt. Rushmore. Today we're driving through the Black Hills which are actually mountains. We're now heading for the Rocky Mountains. Aunt Sylvia says they're even higher. That's hard to imagine.

Love,
Arlene

July 24, Colorado Springs

Dear Mommie,

I'm having a really good time with my cousins, but I miss Puffy and the kittens. Yesterday we looked at the Garden of the Gods. Then we climbed 300 steps to see the Seven Falls, a beautiful spectacle. We had a race up the steps and guess who won?

Love,
Arlene

July 25, Colorado Springs

Don't open! Private unless you are Gertrude Blum. Keep this letter to yourself.

Dear Mommie,

I'm writing this letter in the dark, so I can't see what I'm writing. I'm getting sick of how unfair Aunt Sylvia and Uncle Isadore are. I've been a little angel the whole vacation. Lots of good it does me. Whenever Sylv and Is aren't looking, Stevie hits me hard for no reason. Finally I got fed up and gave it back to him good. Uncle Is bawled me out. Brother, somebody sure is his parents' pet.

Love,
Arlene

July 26, Estes Park

Dear Mommie,

Forget about the letter I wrote yesterday. I was in a bad mood.

Today was the best day ever. We drove up to the snow. I made slush balls and threw them at Stevie and Ronnie. Then I started up a rocky cliff. I thought I'd just go up a little ways, but I kept going. Climbing up was so much fun. Some places were real steep, but I made it. They took a picture of me up high.

When I reached the top of the rock, I stopped and looked down at Aunt Sylvia, Uncle Is, and my cousins way below me. They looked very small and far away.

Love,
Arlene

To the Summit of Denali ▲

1970

EN ROUTE TO HIGH CAMP the next morning, we faced a two-thousand-foot, 45-degree hard-snow slope—the steepest so far. Fortunately for us, it was festooned with red, blue, and yellow polypropylene ropes left by the Japanese. These lines were attached to the slope with aluminum pickets driven into the ice. By clipping into fixed lines for safety, each of us could climb at her own pace.

I clipped one end of a sling of nylon webbing to my waist harnesses and the other end to a jumar ascender. The metal ascender had teeth that allowed it to slide easily up the rope. The jumar stuck if I tried to move it down the rope; similarly, it would catch me if I were to slip and fall. I clipped the jumar ascender onto the blue fixed line and slid it up as far as I could reach, stepped up, and took a deep breath. Continuing to slide, step, and breathe, I established a rhythm, and during the next three hours made steady progress up to the ridge of the West Buttress.

Cresting the ridge at 16,000 feet, I stopped to catch my breath and look around. I was in a world of brilliant light and ice. The summits of Mt. Foraker, Mt. Hunter, and Mt. Huntington shone above a swirling bath of clouds at my feet. But I couldn't admire the view while I was moving, for in places the ridge was little more than a foot wide, dropping off thousands of feet to the glaciers on either side. Climbing the ridge was an extreme meditation, thinking about my breathing and moving with focus, concentration, and harmony. Where I placed my foot determined whether I lived or died. Future plans, past regrets, and the normal clutter of my mind were silenced. I felt a sense of peace and distance from the world reminiscent of that I'd found as a child up in our cherry tree or in the vacant lot watching the clouds.

Margaret Clark climbing the precipitous crest of Denali's West Buttress.

My reverie was interrupted by Grace's and Dana's quarreling voices ahead. I caught up with them to see if I could help.

"My head is pounding and I feel like throwing up," Grace said. "We needed another day to acclimatize. Dana pushed us too fast."

"You can't take altitude," Dana said. "You'd spend a week at every camp and still not be ready to go up."

"I'll stay here as long as it takes me to reach the top," Grace retorted.

I offered them my bag of M&M's. "Thirty days is our limit," I said with a conciliatory smile. "That's all the food we brought."

"If the food runs out, I'll get by on water," Grace said defiantly.

I continued along the ridge past Grace and Dana. But my focus was lost and I began to worry about Grace. I remembered my first climb on Mt. Adams and how Mike was unable to tolerate altitudes higher than 10,000 feet. I wondered if Grace might have her own elevation ceiling—and if she was already above it. As I continued to climb along the elegant narrow crest of the West Buttress, I tried to enjoy the exhilarating surroundings but found myself hoping with every footstep that Grace would adapt to the altitude and that her judgment would not be overwhelmed by her desire to reach the top.

After three hours of steady progress along the ridge we reached our High Camp at 17,200 feet. We found a yellow tent occupied by three Japanese climbers and a small cave carved out of the ice by a previous party. Leaving our loads in the ice cave, we planned to go down to Camp III for the night. "Carry high, sleep low" is the standard rule for good acclimatization.

Grace flopped down on top of her pack. "I'm too tired to move," she said. "I'll stay up here with the Japanese tonight. You two head down."

I looked at the lines of pain and fatigue in her face. "Not a good idea, Grace," I said gently. "You may feel worse in the morning."

"Are you telling me what to do?" Grace glared at me.

"No, Grace, but we need your help tomorrow. Who will carry up the rest of your gear?"

"It's not so much," she said, waving dismissively. "You can each carry a little extra."

"But Grace, who will be the leader if you're up here and we're all down there?"

Grace stared at me with steely blue eyes. "You will."

I winced and felt a nervous flutter in my stomach. I didn't feel ready to be responsible for our team. But I was too intimidated by Grace to challenge her decision to stay, so Dana and I left our loads and Grace at High Camp and headed down as she ordered.

Dana hurried along, and I slowly descended along the crest of the West Buttress alone, enjoying the glorious vistas of the magnificent Alaska Range. I was moved almost to tears by the splendor all around me, and then once again my thoughts turned to Grace. Could it have been altitude sickness—not sexism—that had made her stop short of Denali's summit on her two previous attempts?

The climbing community was all too ready to believe women could not get along. Staying a cohesive team mattered to me as much as climbing the mountain. Faye and the two Margarets had been quietly carrying loads, keeping their distance from the strife in our tent. It was up to me to find a way to deal with the conflict between Grace and Dana.

Down at Camp III the next morning, we awoke to the roar of a savage wind and many feet of new snow weighting down the walls of our tent. There would be no climbing today. The space inside the four-person tent Dana and I occupied was now so diminished that we had to get out of our sleeping bags and dress one at a time. Every half hour one of us headed out into the blizzard to shovel snow off the tent so it wouldn't collapse. I was both worried about Grace and annoyed with her. Why had she stubbornly stayed up at 17,000 feet? There was nothing we could do to help her, so Dana and I settled into a cozy morning of melting snow, nibbling trail mix, and reading. Being able to happily withstand the tedium of bad-weather days is a most useful attribute for a high-altitude climber. My childhood had provided me with extensive practice at dealing with tirades and storms while stuck in a confined space. My best tactic for maintaining equanimity both then and now was losing myself in a good book.

Outside shoveling snow, Mike Bialos yelled across and asked if any of us wanted to come play cards with him and his teammate Jan.

"Sure," I called back. Dana and I floundered through the drifts to Mike and Jan's tent.

"I have to admit we were worried when we heard we'd be sharing the mountain with ladies," Mike said as he dealt the cards to the four of us for a game of hearts. "I guess we weren't especially welcoming."

"Not friendly at all," I agreed. "What was going on with you?"

"It might sound strange, but we were worried your having periods would cause problems," he said, looking a little embarrassed.

"Periods?" I laughed.

"I know it seems silly, but having women on a big mountain like this is such an unknown. And we've put so much into getting here. We were afraid we might have to take care of you if you had cramps or got emotional or something."

"Oh my God!" I couldn't help exclaiming through my bemusement.

"But after watching you climb with those huge loads through the storms this last week, we realized that was ridiculous," Mike continued. "And we're pleased you're backing us up."

"What?" I looked at him quizzically.

Mike explained that the National Park Service required each group on Denali to have a backup in case they got in trouble. Grace and he had agreed that our teams would support each other. He confessed he'd made this agreement with Grace in spite of his misgivings about women on Denali because he had no other good options.

"Grace didn't mention it," I said. "But of course we'll help. And I'm pleased you'll back us up." We shook hands. I was happy that Mike's strong team was on the mountain with us, that they now respected us as climbers, and we were becoming friends.

That afternoon, the Japanese descended from High Camp and Grace wisely came back down with them.

The next day the winds dropped and we all climbed back up the ice slope and along the crest of the West Buttress to High Camp. Our group moved into the cold, clammy snow cave. As I sat on the shiny ice inside the cave melting snow for dinner, I noticed Grace gobbling a handful of pills. I asked her what they were for.

"Diamox and Lasex for altitude, pain pills for migraine, vitamins A, C, and E—" Grace stopped abruptly and covered the remaining pills with her gloved hand. "It's none of your business," she said sharply. "I'm the doctor here, not you."

"You know best," I said, worried, but uncertain how to help Grace.

After dinner I slid my sleeping pad and bag into a shallow alcove in the ice and wriggled into it. But sleep wouldn't come. We were only one day's climb from our goal. But what about Grace? Would I have to send her down? Would I dare stand up to her if necessary?

Tomorrow, the time would come for us—the first group of women ever—to try to reach the highest point in North America.

On July 6, 1970, our summit day, the alarm shrilled at 2:00 A.M. Faye went out and reported that a ground blizzard was blowing hard and her thermometer read minus 20 degrees Fahrenheit. Getting dressed that morning was relatively easy, as I had slept in my regular underwear, my long underwear, a cotton turtleneck, wool sweater, down vest, heavy down parka, wool pants, down pants, two hats, face mask, wool mittens, and two pairs of socks. In the beam of her headlight, Margaret Young melted piles of snow and cooked breakfast. She handed me a mug of oatmeal and I tried to choke it down before it froze. My mind was so fuzzy that lacing my boots was an exacting intellectual exercise. Finally I put on my Windbreaker, wind pants, boots, outer mittens, and daypack and then strapped metal crampons onto my feet. I struggled out of the small, oppressive cave and was shocked awake by the icy air.

We began our climb at 4:00 A.M. in the arctic dawn, following faint scratches in the ice left by Mike Bialos's team, which had gone ahead of us. Our first goal was Denali Pass, nearly a thousand feet up. I established my rhythm as John Hall had taught me years before: Step, breathe, breathe; step, breathe, breathe. Clad in bulky layers of clothing, we moved steadily upward, each of us lost in her own thoughts. A bitter wind froze the vapor in my breath to my parka

At High Camp on Denali, the evening before our summit attempt, 17,200 feet. *(Photo by Faye Kerr)*

hood, forming a grotesque icicle ruff around my face. My fingers, nose, and forehead ached with the cold. When I tried to wiggle my toes, I felt nothing.

The first slanting rays of sunlight failed to ease the bite of the air I inhaled with each rasping breath. Coughs from Margaret Clark just ahead of me punctuated the howl of the wind and my labored breathing. As I kicked my cramponed feet hard into the frozen snow, I tried not to look down at the slope below me. I felt more secure looking up toward the pass ahead, the sky, and the unseen summit.

After nearly three hours, Margaret and I topped Denali Pass and were slapped by a vicious blast of wind. I bent over my ice axe, braced my body with all my strength, and struggled to stay on my feet. After some minutes, the wind eased a little and I looked up and saw a lenticular cloud hovering over the summit like a malevolent flying saucer. Margaret and I fought our way to the lee side of a large rock and crouched down. I took a strip of moose jerky from my pocket, bit down, and nearly broke my tooth. It was frozen solid. I put the jerky inside my shirt to thaw. Faye joined us and we discussed turning back, but the others were ahead. We decided to continue until we caught up with them, hoping conditions would improve.

And in the late morning, they did. The wind dropped, the clouds retreated from the summit, and the sun changed from a pale disk to a fiery ball. The indigo sky contrasted with the snow, immaculate white and sparkling. As we moved steadily upward past outcrops of pink granite and black basalt, the rock crystals glistened in the hard light. Each grain of snow was distinct and perfect. Objects and emotions alike were sharply defined up here.

Wearing brightly colored down clothing, my five companions looked like a string of M&M's moving across the snow. I felt a deep affection for them, and gratitude that together we had reached this extraordinary place. Dana led the way at a good pace. Margaret Clark followed, still carrying her ice hammer in the hope of finding some of the technical ice-climbing she loved. Faye had shed her down jacket and was moving rhythmically upward with a serene smile on her face. Margaret Young had quit searching for photos and was ascending at a steady pace.

As we climbed, my leg ached a little, but I could keep up with the others. Grace lagged behind and we waited as she hurried a dozen steps and then bent over her ice axe, recovering. She caught up with us and then crumpled onto the snow, holding her head, her eyes sunken, her face drawn and despairing.

Faye offered her some hot lemonade and snacks but Grace brushed her away like an annoying mosquito.

"Grace, you don't look so good," Dana said, sounding concerned. "Maybe you should head down and try again another day."

Grace fumbled in her pocket with her gloved hand. She took out an assortment of pills, put them in her mouth, and swallowed them without liquid. She pulled herself slowly to her feet and lurched upward. We were at 19,500 feet, higher than Grace had ever been before.

Slogging across the interminable snow basin leading to the summit ridge, I positioned myself far enough behind Grace so her stops and starts wouldn't influence my pace. I was both concerned for her well-being and irritated with her for not admitting she was sick. I focused on getting to the top so I could go back to my cozy home in Berkeley.

Then the ridge leading the last seven hundred feet to the summit appeared. We were going to make it. Our success should silence the voices disparaging women's ability. And maybe it would muffle the still-painful voices from my childhood as well.

All that mattered now was taking the next step. The wind had blown the snow into delicate ice feathers that crumpled beneath my feet. I counted one hundred steps over and over. I looked up and whooped out loud. The dark triangle of the summit rocks was in sight. But then I saw Grace, ahead of me barely moving, and my elation dissipated. I caught up to walk close behind her in case she needed help.

We continued toward the summit ridge without speaking. As we reached the bottom of the ridge, an airy place, I could look down an astounding three vertical miles to glaciers and tundra below.

Grace suddenly fell onto her knees. I stopped, took out my thermos, and offered her hot lemonade. And as I knelt next to her, Mike Bialos came by to greet us on his way down from the top.

"It's getting late, Grace," Mike said, glancing at his watch. "Come down with me now and I'll climb up again with you tomorrow when you feel better."

Margaret Clark and Faye joined our huddled group. "It's not safe for you to continue, Grace," Margaret said. "Getting up the mountain isn't the whole story. You've got to get down, too."

Margaret took me aside. "Arlene, you're the leader now. Send her down."

"We'll all go back with you now, Grace, and come back up another day," Faye offered.

Dana joined us just in time to hear Faye's generous suggestion. She rolled her eyes at Faye's idea and gave me a pointed look. She, too, wanted me, as the deputy leader, to order Grace to go down with Mike. This was the moment I had dreaded: Grace nearing collapse and everyone looking at me for decisive action.

"Grace, it would be safer for everyone if you went down with Mike now and tried again tomorrow," I said tentatively.

"I have other things to do tomorrow." Grace heaved herself to her feet. "I'm climbing the mountain today." She continued slowly up, ending our discussion.

Mike shook his head, unpacked his emergency sleeping bag, and handed it to Dana. "Keep this in case . . ." He waved and started down. "And good luck."

Above 18,000 feet, a climber's mental capacity is said to be reduced by half. As I followed Grace up the final slopes, I wondered about my own thinking. Why hadn't I ordered her to go down with Mike? I remembered Grace saying life wasn't worth living without Vin. Did she mean that? I felt terrified as I realized Grace was stumbling past her own limitations to the summit of Denali and didn't necessarily care if she lost her life in the process. As Grace careened along, Margaret Clark and Faye caught up and roped her between them for the final climb to the summit, which was now so close that without discussion we agreed that helping her realize her dream was the kindest course of action.

Grace struggled up the last few hundred feet. Looking dazed and unaware of her surroundings, she collapsed in a shallow depression in the snow just ten feet below the summit. Dana, who'd gone ahead and was waiting on the top, came back down and gently helped Grace up the last few steps.

Just below, I stopped to catch my breath and let go of my concern about Grace. We had all worked so long and hard for this moment, and I wanted to savor it. The wind was still and the sun beat down. As we crested the ridge, a sense of completion washed away my fear and exhaustion and I was as happy as I'd ever been. We were the first team of women ever to climb Denali, the Great One.

Chapter 10 ▲

Shoveling Snow
Chicago 1958

One stormy winter day when I am thirteen, I come home after school to an unusually silent house and a note that Grandma is at Aunt Ruth's, my mother is working at the hospital, and Grandpa is shopping. Suddenly ravenous for meat, I grab my babysitting money and sprint to the neighborhood Jewel grocery store. I'm not allowed to cook in Grandma's kosher kitchen, especially when no one is home, and the worst thing I could cook is non-kosher meat. I buy a two-pound nonkosher sirloin steak.

Back home, I stand at the stove watching the steak sizzle in the pan. I breathe in the rich, meaty aroma that fills the air. As my steak cooks, I cut off pink, juicy pieces from the end and pop them in my mouth, savoring each delicious bite. Then my mouth goes dry at the thought of getting caught. After the steak is done, I open the windows for a few minutes, clean the pan, and wipe up each drop of splashed grease, trying to destroy all evidence of my wrongdoing.

I take the illicit meat to my room and hurriedly devour the entire steak. By the time my family arrives home, I have a cramp in my stomach from eating too much too fast and from the terror of getting caught. I hear an uproar in the kitchen.

"Oy, yoi, yoi!" Grandma shouts. "There's been a thief in the house! My frying pan! The drawer under the stove! It's not there!"

Up in my room, I try to make myself very small.

"My God, here it is! With the milk things! My pan! It'll have to be buried for a month!"

I'm shaking with fear.

"Come down here, you worthless child!" Grandma yells from the bottom of the stairs. "You've been cooking behind my back! You good-for-nothing!"

I slink into the kitchen to face all three of my parents berating me at once.

My mother shakes her head sadly. "Arlene, such a bright girl, you should know better."

"You could have burned the house down!" Grandpa yells. "We'd have been out on the street! You're a no-good!"

I've been caught and I have no excuse. I run from the kitchen and out the front door into a blizzard, not stopping to grab my coat. I fall into the snow and sob. When my tears are spent, I see a snow shovel leaning against the porch. I pick it up and begin to clear the sidewalk of the new snow. As the cold air flows over my hot face, I grow calm and focus on my task.

"Come in," Grandpa yells out the front door. "It's too cold for you out there."

"I'm shoveling the sidewalk."

"At least come in and get your gloves and jacket."

"I'm warm enough."

"You'll get sick out there half naked."

"I'm fine. Leave me alone."

It's true. As I shovel the heavy snow, I admire the beauty of the flakes pelting down and glory in the elemental power of the storm. I welcome the sting of the wind and the driving snow, the feeling of using my body to move the piles of snow. The natural world of snow and storm outside our house is beautiful and strong and good and I am part of it.

Finished with our sidewalk, I move on to the neighboring houses and continue shoveling.

Out in the Cold ▲

1970

ON THE SUMMIT, we celebrated with a feast of chocolate fudge, smoked oysters, and dried apricots hoarded for this moment.

"I always knew we could make it," Dana said. She then glanced at Grace, who was slumped down, seemingly oblivious to our celebration. "Or at least some of us could."

Faye's tanned face looked radiant. "It's lovely up here. I wish we could stay on top for the sunset."

I looked fondly at my companions as they relaxed in the warm sunlight. In spite of Grace's altitude illness, Margaret Clark's cough, and my recently broken leg, we had all reached the summit together. For me the question about whether women could climb the world's toughest mountains had been answered.

Looking around, I saw that the nearby mountains were enveloped in cloud up to 12,000 feet, leaving only their summits poking through. A storm was probably brewing in the cloud layer directly below us, but here all was calm.

"Time to line up." I tried to organize the others for a group photo. "This is your chance to be a cover girl."

Everyone posed around Grace, who was ashen and unmoving. I took photos of the team in their colorful hooded down parkas on the highest point in North America. It was 5:00 P.M. and there was little time for congratulations. We needed to get Grace down, fast.

Helping Grace along the summit ridge and then across the enormous, nearly flat plateau below was our first challenge. First Margaret and Faye, then Dana and I, supported her in the track we had broken coming up. We tried to move quickly, but with Grace's weight, we sank at every step. She was in a stupor, staggering between us. At the ridge of the Archdeacon's Tower at 19,600 feet, Grace fell down onto the snow, vomiting and unable to move any farther. She refused food or drink, moaning, "I'm going to die here. Leave me in peace."

For a moment I was tempted to do as she asked and leave her. I was fed up with her stubbornness, her grief, and her inability to consider the rest of us. It would serve her right if we went down without her.

But of course we couldn't. I realized with horror that all of us, not just Grace, were in a desperate situation. Exhausted from our long ascent, three thousand feet above our camp and supplies, we had to get Grace down the mountain or else stay up here with her. We needed a leader with a plan of action. Without time for deliberation I, at age twenty-five, had to become an expeditionary leader in a life-or-death situation.

Denali photo © Bradford Washburn, Courtesy Panopticon Gallery, Waltham, Massachusetts

I closed my eyes and took a series of slow breaths. "Let's all take a break," I suggested, trying to sound decisive and keep my fear at a safe distance. "That'll give Grace time to rest while we make a plan." Sitting on our packs, nibbling on frozen candy bars, we brainstormed.

"I'll stay here with Grace until she's better," Faye said. "The rest of you go down."

"Not safe," I objected. "We're near where the Wilcox party ate it—and they had sleeping bags. That storm could blast in anytime."

"Give Grace some Dexedrine," Dana suggested. "That'll perk her up."

"She's already drugged," Margaret Clark said. "Adding Dexedrine to the mix is a bad idea. When it wears off, she'll be worse than ever."

"Besides, she couldn't keep it down," I added.

"Go on . . ." Grace whispered so softly we had to lean in close to hear her. "Leave me . . . in peace."

"One of us can catch up with the Bialos team," I suggested. "Maybe they can bring a tent back up to Denali Pass. We can get her down that far."

I looked up and saw to my surprise that there were only four of us gathered around Grace, who was slipping in and out of consciousness. Margaret Young had headed down and was disappearing from sight.

"I expect she's going for help," I said. She'd left without a word.

We had no idea if Margaret could catch up with Mike's group, or if anyone would be able to get back up to us. A storm was poised below. It was getting colder by the minute. We were too high for a helicopter. "Listen," I said, looking each woman in the eye. "We can't count on Margaret or the Seattle team or anyone else to rescue us. It's the four of us and we've got to figure out how to get Grace down."

"I made a stretcher from a pack frame and a climbing rope once," Margaret Clark said.

"We did that in Peru, too," I said. We had a rope and Dana, fortunately, had carried her pack frame up. I tried to sound confident, but my hands were shaking. For the four of us to lower a semiconscious person down the mountain sounded impossible, but we had no choice.

We positioned Grace's immobile body in Mike's sleeping bag, laid her on Dana's pack frame, and padded her with our spare clothes. Under Margaret Clark's tutelage, we wove the climbing rope around her body and the pack frame to fashion a makeshift rescue litter. My fingers ached as I clumsily tied knots in the rope. Finally Grace was a manageable bundle on the pack frame, encased in a cocoon of sleeping bag and woven rope. We attached nylon webbing slings on the front, back, and either side of the makeshift stretcher so that we could control her descent. It just had to work.

But when we tried to slide the stretcher down the slope, it wouldn't budge. The snow surface was too soft.

"Maybe... if we hold some of her weight off the snow," I suggested. "I'll lift from this side and Faye, you take the other." Faye and I pulled up on the slings attached to either side of the stretcher. Margaret Clark guided it from the front and Dana pulled up and pushed from the back. The stretcher shuddered and then began moving slowly. We let out a cheer.

For several hundred yards the stretcher slid easily down the glacier, but when the slope became steeper, the sled started going faster and faster.

"Everybody. Move back here! We've got to slow it down," I shouted, and moved behind the stretcher. We wrapped the nylon slings around our waists and leaned back to slow the stretcher.

A crevasse showing green ice stopped us. "Move to the uphill side of the stretcher," I yelled. We had to traverse the steep slopes above the gaping crevasses while preventing the makeshift sled from slipping sideways and toppling over. By lifting or anchoring as the slope demanded, we worked out a system. I told the others when to change sides, lift, or break. Everyone did what I said mechanically. There was no time for questions or exhaustion or fear.

"The rope is choking me," Grace moaned, returning to consciousness. We loosened the rope, but then she started to slide out of the stretcher. Trying to hold her in the stretcher securely without hurting her, we had to stop frequently and adjust the ropes to keep the tension just right. My fingers throbbed each time I stripped off my gloves to retie the knots.

Descending a steep, icy stretch, I felt a sharp pull on the sling as someone slipped. As I grasped the sling and struggled to keep my balance, I realized we could all die at any moment.

As we reached the sunny southern slope, the surface became softer and we sank to our calves in heavy snow. My body pleaded for a rest and I could see how tired the others were. But we had to continue.

Finally the level snowfield of Denali Pass came into view below. It was 11:00 P.M.; the light was fading and the temperature dropping fast. We had been up for twenty-one hours. We were at 18,600 feet, about three hundred feet above the pass, but the slope was steep and freezing hard. We could no longer control the descent of the stretcher. We stopped.

"I'll go down to make sure we get help," Dana said.

"My thermometer reads minus eighteen degrees Fahrenheit and I can barely feel my fingers," Margaret Clark said, waving her arms to keep warm. "With my circulation, I need to keep moving."

"You can all go down. I'll stay here with Grace until she gets her strength back," Faye offered.

They looked at me to see what I was going to do.

I hesitated. The thought of heading down to the world of living creatures with the others was tempting. But I couldn't leave Faye up here alone with Grace.

I decided I would stay up high with them. This was the only choice I could live with—but I hoped I wouldn't die from it.

As Margaret and Dana disappeared, Faye and I finished chopping a level platform in the ice in the lee of a large gray rock and put Grace between us. Cold, tired, hungry, and thirsty, I huddled, shivering, next to Grace's prone body.

"What a gorgeous sunset," Faye exclaimed. I looked up to see the tips of the peaks shining with the golden light of the setting sun, swimming in a sea of orange, magenta, and scarlet-colored clouds. It was indeed glorious but I was more focused on my numb toes. My socks had gotten soaking wet inside my vapor-barrier boots and now both my socks and feet were beginning to freeze. Faye gave me her down mittens. I took off my socks, rubbed my feet, and put the mittens over them. I wiggled my toes: "Wiggle, two, three. The bear went over the mountain. Wiggle, two, three, to see what he could see." The arctic sun dropped below the horizon, leaving magenta and orange stripes across the sky.

Then it was truly dark, and 30 degrees below zero. Would anyone come back up to help us? It seemed impossible for Faye and me to get Grace down by ourselves. If the storm swept in, would we descend, leaving Grace to die alone, or would we stay and die with her?

As I shivered and wriggled my toes, I replayed the last few days in my mind. Until now, I had no personal experience with death in the mountains and it hadn't seemed a real possibility to me. Grace's depression from losing Vin and her inability to adapt to altitude should have been obvious. Why hadn't I insisted that she turn back?

I had expected Grace to be a strong high-altitude climber and had overlooked all evidence to the contrary. I had been so focused on attaining something I wanted that I had not asked enough questions to learn the reality of the situation with Grace. On the other hand, maybe I shouldn't be so hard on myself. She probably would have ignored me and continued up no matter what I had said or done.

As the freezing night dragged on, Faye and I retreated into our own thoughts. I couldn't help smiling as I remembered my family's reactions to a *San Francisco Chronicle* article including photos of Dana and me rock climbing:

"Oy. Your engagement picture. That's what I want to see in the paper. Not you hanging on a rope off some rock," my grandmother pronounced.

"Such a beautiful picture," my mother gushed. "But you are too intelligent for such a coarse activity. I cannot imagine how you can go to the toilet."

"For this we took care of you all those years," my grandfather had chimed in. "Climbing is for rich goyim, not for you. Too dangerous."

I began to shiver violently and tried to calm myself by thinking of John Hall. He'd been selected by NASA to be mission commander of a sixty-day simulated space flight while we were on Denali. John was currently living in a sealed-off chamber. Was it this cold in simulated space?

I dozed fitfully and dreamed that I was curled safe in my comfortable bed, my two cats purring next to me. We were warm and cozy except someone had left a window open and the snow was blowing into the room. Mike Bialos was there telling me to get up and close the window.

Then I opened my eyes to see Mike Bialos standing in front of me, and Margaret Clark's slight form next to him. They were real. Incredibly, they had found the strength to climb back up to help us.

"Thank goodness!" I exclaimed, groping for my stiff socks and boots. "How did you ever make it up here? And Margaret? Unbelievable."

"Good to see you both," Mike said as he helped me to my feet. "Dana and two of the guys are pitching the Japanese tent at the pass. It's just four hundred feet or so down."

"How did you find us in the dark?"

"I showed him," Margaret spoke up. "But I'm freezing. We need to get down."

I looked toward the pass below and in the first, flat light before the dawn saw a soup of thick black clouds boiling below us.

"The slope is steep and hard with lots of little crevasses underneath," Margaret said. "Lowering the stretcher isn't going to work."

"Maybe we can wake Grace." I peeled back the sleeping bag. Grace was conscious.

"Grace, can you walk a short way down?"

I thought I saw her nod. Mike and I hauled Grace to her feet. She sagged limply against us, but with our semi-carrying her, Grace began moving down the slope. Faye and Margaret belayed us from behind. I was so focused on keeping Grace upright that I didn't think about where I was placing my own feet. Suddenly the crampon came loose from my right boot and I lost my footing. Quickly shifting all my weight to my left foot and waving my arms, I managed to regain my balance. As Mike supported Grace, I whipped off my mittens and fitted the metal crampon back onto my boot with my bare fingers. My fingertips froze within seconds. Even though I immediately put my mittens back on, stabbing pain radiated up from my fingers as I laced the crampon straps around my boot.

It seemed an eternity before we reached the safety of the orange tent at Denali Pass. We helped Grace into the small tent and then into a sleeping bag. Dana and I crowded in along with Steve and Jan, the other climbers from Mike's party who'd brought the tent back up.

"I think I'll sleep outside," Faye called into the tent.

"In the snow without a sleeping bag?" It was 20 below zero and blowing hard.

"It's lovely out here," Faye said. "Wearing my big down jacket and pants, it's like the Ritz compared to being the sixth person in a two-person tent."

Mike and Faye made another trip all the way back up to our bivouac site to retrieve the gear we had left there. And Margaret Clark somehow managed to descend alone to the snow cave at 17,200 feet.

Wearing all my clothing, I lay on my side, trying not to touch the icy tent, and fell asleep. A couple hours later, I awoke to the sound of Grace futilely retching.

Grace was able to walk with some help in the morning. As the others guided her down the track to High Camp, I stayed behind to pack the rest of our gear into one massive load. After descending the steep slope below the pass, I was so exhausted that I could barely walk with my burden. I straggled across a level basin through deep snow, counting my breaths and willing myself to take a step each time I reached three. "One, two, three. Lift that foot and put it down. Keep going. Don't stop. You can do it." I wanted to sink down into the snow and go to sleep. Finally I climbed up the small rise to High Camp, dumped my load, and crawled inside the ice cave. Margaret Young welcomed me with a cup of hot cocoa. Compared with where I'd spent the night, the dank ice cave at 17,200 feet seemed a sultry haven.

A few minutes later, three of the Seattle climbers yelled into our cave that a severe storm was forecast and they were heading down to a lower camp.

Although Grace was sitting up and sipping hot soup, we were still in danger. Realizing we needed to rest, eat, and drink before we could safely descend the narrow ridge, we packed for an early morning departure and fell into an exhausted sleep.

We awoke to see fresh snow filling the tunnel to our cave, and from the doorway could hear the vicious howl of the wind. We had no choice but to descend. The gale drove stinging ice crystals into our faces as the five of us packed huge loads down the narrow ridge of the

West Buttress. I roped Grace just in front of me so I could catch her if she fell. Picking my way along the precipitous crest in a cloud of snow, I felt like a blind tightrope walker in a wind tunnel.

When we reached the top of the steep ice face, the wind dropped suddenly and the clouds rolled away, uncovering dazzling views of the Alaska Range in every direction. The sky was navy blue and the newly fallen snow was brilliant. The change was so unexpected that I felt a surge of euphoria. We were all going to get down alive.

Margaret Clark and I began snapping photos of our descent. Unclipping from the fixed line, we backed off to the side to take a dramatic profile of our team descending the steepest part of the face.

"Get back on the line," Grace ordered us from across the slope. "It's dangerous to unclip. I'm the leader of this expedition and if anything happens to you, I'm responsible."

I heard Margaret Clark's cynical laughter behind me.

I felt a burst of anger at Grace, but told myself I should be relieved that she was recovered enough to shout orders at us. I finished taking the photos and, without a word, clipped back into the fixed line.

We snowshoed down to our tents at Camp III, singing children's songs at top volume. A warm sun shone down on us. We were three days above Base Camp.

I dug up a cache of food we had left behind and gave Mike Bialos a bag of Ghirardelli candy bars for his team. "Thank you," I said with a big smile. "Without your help, who knows what would have happened."

"You'd already hauled her most of the way," he said. "We just helped the last little bit."

"A vital bit," I said, giving him a big bag of cashews and a grateful hug. He then told me he had made a third trip up to the bivy site that night and slept up there. I hadn't thought about where Mike had spent the night of the rescue, and was awed by his stamina.

The next morning the storm returned and we continued down, plagued by soft snow and a stinging wind. At Camp II, our tents inflated like willful orange balloons as we fought to pitch them in the blizzard. When they were subdued and firmly in place, we crawled inside. Dana and Grace pointedly moved into different tents.

For the next four days and nights, we struggled to keep our tents from blowing away in the gale. The third night, when the storm was at its most violent, we took turns holding on to the center pole all night to keep the tent upright.

Despite the storm, Grace appeared to recover and was eating her dried moose jerky with a hearty appetite. Staying to herself in the tent she shared with Faye and Margaret Clark, she appeared to remember little about the summit day and talked as though she had been strong the whole time. At first we didn't contradict her.

"Three of my carabiners and two slings are missing since the summit," Grace grumbled one evening when all six of us were cozily squeezed into one tent for dinner. "Somebody took them and I want them back." She fixed her gaze on Dana.

"Nobody took your damn carabiners, Grace. And you're lucky all you lost was some climbing gear," Dana exploded. "You could have died up there and taken some of us with you."

Looking down at my fingertips, purple with frost from my night out with Grace, I nodded my agreement.

"I would have gotten down the mountain just fine if you had just given me a swift kick. Men would have done that." Grace's face was ashen with rage. "You tied the ropes too tight in that miserable stretcher and hurt my leg. You were behaving like hysterical women."

Margaret Clark joined the fray. "We saved your life, Grace, and you risked ours."

"Grace, you asked us to leave you on top, don't you remember?" I said, trying to stay calm. "We almost died trying to get you down. We'd appreciate some thanks."

"You made up this rescue drama," Grace said, now shaking with anger. "I'm going to have to miss work while my leg heals. I could sue you for my lost wages."

For a wrathful moment, I wished we had left her on top. I wanted to tell her what I thought about her risking her life and ours, too, but I decided an open conflict would not help us outlast the storm. I calmed myself by thinking that it was not unusual for rescuers to get reproach rather than thanks from those they save. And it wasn't clear that Grace had wanted us to bring her down the mountain and save her life.

Letting go of our disappointment at Grace's lack of gratitude, we tried to make the long storm days as pleasant as possible. While the wind rattled our tents and the snow blanketed down, I made popcorn and Margaret Clark shared her stories of climbing in warm, sunny Morocco.

Finally, on the fifth morning, we awoke to still air and a cloudless blue sky. We packed up and hurried down the glacier in the blazing sun—as much as one can hurry through more than two feet of wet, heavy snow. Led by the ever-energetic Dana, we made it all the way to Base Camp and the landing strip by nine that night. Keeping our gear packed for an early morning departure, we slept on the snow without tents.

"It's hard to believe," I said when we were lying in our sleeping bags looking at the darkening sky with the white bulk of the mountain in the distance. "Here we are just below the Arctic Circle, sleeping on the snow at 7,200 feet, and it feels like the tropics."

Margaret Young passed around a bag of dried pears. Off to the side, Grace read a book of Alaskan poetry by flashlight. Looking up at the stars, I relived the last few days. Although I was not very fast, I had succeeded in reaching the top of Denali just six months after a serious break in my leg. I felt I had done a good job as a participatory leader, listening, supporting, and encouraging the team. I did less well when situations required a more authoritarian style of leadership. I had been unable to send Grace back down when she was ill—an almost fatal mistake. But I was happy with myself for organizing her rescue and then choosing to stay with her.

The next morning, we were awakened by a distant hum, and a few minutes later Don Sheldon landed his Cessna. Dana, Margaret Young, and I threw our bags into the plane and clambered aboard. The others would follow shortly. Flying out amid the vast glaciers of the Denali Range, I felt centered, serene, and reluctant to leave this splendid domain of ice and snow.

The local newspapers in California covered our expedition, but the national papers and climbing community ignored it. It was too early for the idea of an all-women's team to capture public attention. Indeed, some years later I read a newspaper article about another team of women attempting to make what the paper called the "first all-female ascent of Denali."

In April of the year following our climb, I was saddened to read in the newspaper that Grace Hoeman had died in an avalanche while descending from a hut near Anchorage during a snowstorm.

Years later I read Grace's obituary in the *American Alpine Club Journal,* written by an Alaskan physician friend of hers. My attention was riveted when, after referring to Vin's death on Dhaulagiri, the author wrote that Grace met "a not unwanted similar death." I once again felt grateful that her death had not come near the top of Denali, and that she hadn't taken us with her.

I went on to read that she had received her bachelor of medicine degree from the University of Berlin in 1944, at the end of the Nazi regime. I was stunned, and wondered if living in the Nazi capital during those desperate days might have contributed to her tolerance for suffering, adversity, and risk. I read that she had survived two bouts of tuberculosis, which could explain her difficulty adjusting to high altitude. Grace had two daughters and three grandchildren, none of whom she ever mentioned to us. The obituary ended by saying that Grace was an unusual and extraordinarily gifted person. I wished I'd gotten to know her under different circumstances. There were so many questions I could have asked Grace during those long tent-bound storm days, but the answers would never be known to me.

It wasn't until almost thirty years later that I learned just how notable our rescue of Grace was. In the spring of 1997, the Mazamas, a mountaineering club in Portland, Oregon, invited me to speak at a conference on climbing Denali. After I told the story of Grace's rescue in public for the first time, Vern Tejas, the head guide at Denali Park and a legendary climber, came up to congratulate me.

"Rescuing Grace was quite an accomplishment," Vern said. "I've been working on Denali for years and haven't heard a story like yours that ended happily."

I appreciated his words, especially because I'd just seen his slide show of the first winter solo ascent of Denali. Vern had described climbing in the darkness carrying long aluminum ladders horizontally over each shoulder to stop him from dropping into hidden crevasses.

I told him I found his winter climb remarkable.

"Not nearly as tough as getting a comatose person down Denali. Three of us were on the summit, two clients and me. One client became unconscious." He looked down at his feet. "A storm was coming in. I tried to carry her down, but she was too heavy . . . I had to save the other client. There was no choice but to leave her and go down. It was the worst thing that ever happened to me."

"I'm so sorry," I said, once again thankful that we hadn't had to face such a terrible choice.

"People who've passed out on top of Denali have died there as far as I know," Vern said. "You should be very proud of saving Grace."

Vern's words, so many years later, made me look at our Denali climb in a different light. I'd had no idea that our getting Grace down safely was so unusual. I now appreciated that my being forced to take charge of her rescue had given me leadership experience and confidence at a young age and had launched me into adventures I never could have imagined.

Chapter 11 ▲

Strawberries

Excelsior Springs, Missouri, March 1, 1951

I stare at the bowl of mashed strawberries floating in evaporated milk. I'm six years old today.

"Eat," Grandma commands.

I try a little. The milk tastes weird. I push the bowl away.

"What's wrong with you? They're delicious. And not cheap." Grandma puts the bowl back in front of me. "They're special for your birthday. Eat."

"I want my mommy." This has been my refrain ever since last fall when we left Iowa to live in Excelsior Springs, Missouri.

"You know Gert's in Chicago. Working to support you," Grandma barks. "Eat."

Grandma returns to the TV. Grandpa, smelling clean, and just back from bathing in the healing waters of the springs, lifts me into his lap.

"I miss Mommy." My eyes water.

Grandpa gives me a comforting hug. "Your mommy misses you, too, but she has to work."

"Why can't she work here?"

"In Chicago she'll find you a daddy."

How do you find a daddy? Maybe you need to pay really good attention, like when you want to find a shiny coin in the street. Or maybe you need to know some magic words.

"Why can't she find me a daddy here?"

"Chicago's a big city," says Grandpa. "Lots of Jewish men there."

"More daddies?"

"Yes, more daddies. Now, what about your snack?"

I can't eat the strawberries in the strange evaporated milk, but manage to choke down a piece of the chocolate birthday cake Grandpa baked for me.

Mommy's gone the whole year I'm in kindergarten. It must be my fault she's away. I've done lots of bad things. Like not eating my strawberries.

Avalanches ▲

1971

THE ICE WORLD OF DENALI receded into memory as I toiled away, trying to finish the research for my PhD. One foggy evening, I worked through dinner in my lab purifying tRNA for my final experiments. My fellow students were long gone and the lab was eerily quiet except for the repetitive ticking of the spinning centrifuge. Above my lab bench, a huge banner proclaimed DO IT in foot-high letters. But after Denali, I was finding indoor experimental work tedious. And I was lonely. After my boyfriend, Howard Simon, had left for graduate school in San Diego, we had drifted apart. John and I hadn't been in touch for a long time.

Just as I finished purifying the tRNA, I knocked the vial onto the floor, where it shattered, the precious liquid lost. Needing a break, I headed across campus to the Northside Theater. I was just in time for a late showing of *The Endless Summer.* Watching two tanned young California surfers frolicking on sunny beaches on every continent searching for the perfect wave, I decided that an ideal reward for completing my PhD would be my own alpine version of their quest: an *Endless Winter* of climbing around the world in search of the perfect mountain.

Returning from the movie to the chemistry library, I happened upon a world globe. Spinning the globe, I realized that by traveling north and south across the equator for a year and a half, I could climb in the world's highest mountain ranges during the best weather. Beginning with autumn ascents in the European Alps, I could climb during the winter in Africa, spring in Iran and Kashmir, summer in Afghanistan, fall in Nepal, stop at Australia's Great Barrier Reef for R&R, enjoy another "summer" in New Zealand during our winter, and finish off in the South Pacific, exploring mountains under the sea with scuba gear. The more I studied the globe, the more sense it made. Traditionally, climbers flew halfway around the world for a single expedition. Doing a dozen consecutive climbs would be an economical way to visit many of the world's major mountain ranges during one trip. As our strength, skill, and acclimatization increased, we could climb peaks of ever-increasing difficulty and elevation.

Climbing Denali had given me the confidence to dream. And once I could picture the Endless Winter, I was optimistic that I could make it happen. All I needed were companions, money, eighteen months, and a finished thesis.

At Indian Rock a few days later, I met Joel Bown, a muscular young man with short curly hair, an affable grin, and very thick glasses. As we made several tough ascents, I appreciated Joel's climbing skill, equanimity, and sense of humor. Afterward, we relaxed on top of the rock, the sail-dotted San Francisco Bay spread below us. Joel told me he was visiting Berkeley from Salt Lake City, Utah, where he was a computer programmer. He'd been saving money for a climbing trip to Europe that summer, but the charter company with whom he'd reserved his flight had gone bankrupt. He was looking for another adventure. Feeling an instant connection with Joel, I told him about the Endless Winter and invited him to climb with me for a year and a half. He thought for only a moment before he said, "Sure, why not?"

I was elated. I had my first recruit. Wanting a team of four, if possible, I put a classified ad in *Summit* magazine. A little later I received a special-delivery letter from a mysterious Colonel Walter E. Traprock, LSD, BS, CCC, applying for a position on my team. He wrote suggesting that we could enhance our commercial appeal by hiring Playboy Bunnies to ride camels across the Afghan desert, and document our adventures in a book titled *In Search of the Snowflake, or Oblivion Beneath My Boots.* The return address on the envelope was John Hall's PO box at Caltech. The long silence between us was over.

And then the next weekend Colonel Traprock himself came to Berkeley and pulled me back into his life with an invitation for an audacious venture in Canada's Yukon Territory. John wanted to attempt the first-ever traverse across the summits of the second- and fourth-highest peaks on the continent, Mt. Logan (19,850) and Mt. St. Elias (18,008). These mountains were notorious for both avalanche danger and severe weather. I knew and liked the other team members—Susan Deery, my cheerful Reed roommate and climbing partner during the wonderful summer after my senior year; Lucille Borgen Adamson, another student who helped teach the Reed climbing class; and Lucille's husband, Stanley Adamson. The three were graduate and postdoctoral students at the world-class Institute of Molecular Biology at the University of Oregon. Lucille and Stan were famous for keeping unusual pets, including two lively Norwegian elkhounds and a full-grown puma.

My fantasies about John rekindled; I was enticed by his invitation. We would finish our PhD theses, I would join his Yukon expedition, and he would come on the Endless Winter. Ignoring our history, I continued to dream that if I could just be strong and smart and thin and pretty enough, John would fall in love with me and we would live happily ever after.

Meanwhile, the Endless Winter team was growing. Toby Wheeler, a student at the University of Alaska, came up to talk with me after a Denali slide show and signed on. Margaret Young from Denali planned to meet us for a month in Afghanistan. Dave Graber, an easygoing premed student at UC Berkeley, responded to my notice on a bulletin board at the UC Hiking Club and decided to join us in Asia after graduation.

With a core team of four—Joel, Toby, Dave, and me—my dream was becoming real. Going around the world for eighteen months with three guys I had just met didn't seem unreasonable to me; I assumed that any climber who shared my enthusiasm for the Endless Winter would soon become a good friend.

I thought the trip an ideal break between graduate school and my postdoc. My family did not agree. Grandpa told me to use my PhD to earn some money instead of wasting it on climbing and Grandma warned me that all the nice Jewish boys were getting snapped up. I didn't pay much attention to their objections, but my grandfather was partially right: I didn't have enough money. For years I'd been saving what I could from my graduate stipend of $212 per month in hopes of being invited on an expedition, but I'd spent a good chunk of my savings for Denali. Now I scrutinized every spending decision, used a bicycle rather than a car, stopped making trips to Chicago to visit my family, and cut back on long-distance phone calls. When *Summit* magazine offered $100 each for a series of articles about our adventures, I could just manage to pay for the trip.

Motivated by my desire to join John's Logan–St. Elias traverse and to begin the Endless Winter, I managed to complete my thesis research. My experiments verified the H model of tRNA. During the intervening four years, other people had suggested and published numerous different models. As I surveyed the literature in preparation for writing my thesis, I was pleased to see that most of the other experimental evidence fit the H model.*

Fearing I would lose interest in chemistry during my time away, my advisor, Nacho, was not pleased about the Endless Winter. But after I was awarded a National Institutes of Health fellowship for postdoctoral research at the University of Washington in Seattle, Nacho gave his reluctant approval. His daughter Kathy made me a flag with *tRNA* stitched in bright colors. I would fly the flag, I assured Nacho, from all my mountain summits—and think about my research.

One Friday afternoon in March 1971, I received a letter from Aunt Ruth saying Grandpa was ill and I should come home. I dialed the familiar number and the line was busy. I hung up, tried again, and then sat there, sick with worry. Grandpa couldn't be that sick or surely someone would have told me sooner. I dialed the airline, and as I waited on hold, I thought about Grandpa and how much I loved him. He had been my most stable and encouraging parent, a buffer between me and Grandma's wrath. He taught me the Hebrew prayers. He believed in me. I booked a flight for the next morning and redialed our home in Chicago.

"Hi, honey," Aunt Ruth answered in a faint voice.

---

*A few years later, Alex Rich at MIT determined a crystal structure for tRNA, which gave the precise location of every atom. He found that the three-dimensional structure was an H folded into an L shape, similar to what I had predicted from my model based on stringing the colored seed beads from Telegraph Avenue. In 1982, Aaron Klug was awarded a Nobel Prize for similar work on the structure of tRNA and on other biologically important nucleic acids.

"I'm coming home tomorrow morning," I said. "How's Grandpa?"

"I'm so sorry, Arlene," said Aunt Ruth. "He died last night. His funeral was this afternoon."

"No! Grandpa! No!" I was in anguish. "Why didn't you call me?"

"It all happened so fast," Aunt Ruth said, and told me my grandfather had died from a sudden heart attack on Thursday evening. According to Jewish custom they had the funeral on Friday before Shabbat. I hardly heard a word she said.

"Why didn't you tell me sooner he was sick?" I cried into the phone.

"You were far away and focused on your own life," Aunt Ruth said. Then, saying it was not a good time to talk, she hung up.

Grandpa was dead. I never got to say good-bye. I never got to tell him how much I loved him. They didn't tell me. I hated them. I hated myself.

I wept until I had no more tears, stumbled into my room, and slept. I awoke with a furry feeling inside my mouth and reconsidered my flight to Chicago. Why face my family now? With Grandpa's funeral over, there was little reason to go home. I went back to sleep.

The next day I went to my lab, but could only sit and cry. If I hadn't been so intent on my research and my climbing, I'd have called home more and found out Grandpa was ill. If I hadn't been saving money for the Endless Winter, I'd have gone home and seen him one more time.

I needed to write my dissertation, but instead wrote long letters to my family, trying to accept Grandpa's death. But every time I saw an elderly man on the street, tears would blur my eyes.

For two painful months I couldn't focus on my writing; joining John's Logan and St. Elias trip was now out of the question. Toby Wheeler, who was waiting for me to finish my thesis so we could begin the Endless Winter, decided to climb in the Yukon with John and friends while I worked away.

I had just begun writing again when Aunt Ruth called with more terrible news. There had been a fire in the apartment where my mom and Grandma lived and they suspected the fire was caused by my mother's cigarettes. Ruth told me they'd found a place for Grandma in a nursing home, and my mother was coming to live with me. She was too unstable to live on her own. There was no other option, Aunt Ruth said, it was a daughter's responsibility.

"Give me time to make a plan. Don't send her right away," I begged Ruth.

With shaking hands, I dialed Aunt Sylvia and asked if my mother could stay with her for a while.

"It's your duty to take care of your mother," Sylvia echoed Ruth. "We've bought her a one-way ticket to San Francisco. She's arriving Tuesday."

"No, no, you can't do that. Why can't she get her own place in Chicago?"

"She can't live alone safely," Aunt Sylvia said. "She'll die if you don't take care of her."

"You don't know that. She's just fifty-four. I can't do it. I have to finish school."

"You can get a job and support your mother," she said. "It's time for you to stop being so selfish."

I'm not the only one who's being selfish, I thought as I put the phone down hard.

Although I'd not spent much time with my mother in the nine years since I'd left home, I knew she had remained dependent on my grandparents. Now that she had lost both her father and her home, she apparently wanted someone else to take care of her. I was happy to have her visit me and take a break from her problems. When I phoned her and offered to come to Chicago and help her find an apartment, she said she had to live in California. And then a few days later, she arrived at my front door in a cloud of smoke.

"Arlene, I'm here to take care of you," she said, giving me a hug and a kiss. "I'm so proud of you. You're the most wonderful daughter in the world. But you work too hard and you're too refined to do this crazy mountain climbing. Grandma always says that a woman needs a husband to take care of her. And too much education is bad for a girl. Family is important. I need family. And you're my family."

She continued her monologue as she unpacked. "Can you make me a little something for lunch? Anything will do . . . That's so nice of you, but I never eat bread with seeds and I don't like that tasteless white stuff. Tofu, do you call it? No coffee in this house? Oh, my God. Cats! Get them out! They have to stay outside."

The next day, my mother was in an uproar, demanding to know who stole her cigarettes. My housemate Dee pulled me into her room. "You've got to do something about your mother. We cannot have her sitting smoking in the middle of our living room all day."

"But what can I do?" I asked. "She's my mom."

Dee advised me to talk to a therapist. I made an appointment with a counselor on campus and a few days later, my mother and I went to her office. First she asked my mother why she'd come to California.

"To take care of my daughter," my mother answered. "She's the most wonderful girl in the world . . . but why does she do this crazy mountain stuff? My father always said that climbing was for millionaires and lunatics. She should get these crackpot ideas out of her head. My father always said that children are precious and she is all I have in the world. She should have respect for her own life. I'll help her."

I focused on the smooth feel of the maroon velour couch and tried to shut out my mother's soliloquy.

The therapist gently interrupted the flood of words and suggested my mom ask me how I felt.

"She's too young to live alone," my mother went on, undeterred. "Her climbing is just as nutty as saying I had anything to do with that fire. There was no fire, just smoke. It was the firemen who sprayed water all over and wrecked everything."

"You need to stop talking and listen to your daughter," the therapist said firmly. That, I guessed, was my cue.

I took a deep breath. "I love you, Mom, and I'm sorry your life is so hard right now. I miss Grandpa, too. But you can't live with me for the rest of your life. I'll come to Chicago and help you find a place to live. I know you can take care of yourself."

"In Chicago they'll put me in an old people's home," my mother said. "I can't stand sickness. I'll die if I have to live surrounded by sick people. I need to stay in California. I need to live with Arlene; I have no one else in the world. I'm here to help her."

"Mom, you can help me most by letting me live my own life," I said. "You have to take care of yourself, here or in Chicago."

"Listen to Arlene," the therapist said. "Do you want to help her or hurt her?"

"A mother . . . hurt her daughter?" my mother repeated slowly. "Of course not."

There was a long silence. My hands stopped moving and I held my breath as my mother considered the possibilities. I felt I was in a life-or-death situation, not dissimilar to my toughest moments in the mountains.

"I only want what's best for Arlene. I can't live here by myself," my mother said, her eyes closed. "If I can't live with Arlene, I'll go back to Chicago."

I let out my breath. Although I loved my mother, and was touched by her loneliness and pain, I felt my life was saved.

I realized that her agreement to live on her own—needy and dependent as she was—showed that she did care for me. My mother returned to Chicago and I began to heal.

In late June, John's expedition left for Mt. Logan in high spirits. A few weeks later, I received a happy letter from John, carried out by another climber. They had traversed Logan successfully and were en route to St. Elias. Best of all, John wrote how much he was looking forward to seeing me in August and would try to join the Endless Winter. Because my thesis was taking longer than expected, the plan now was to skip Europe and fly directly to Africa in the late fall.

As I worked away on my thesis, tender images of John floated into my mind. Sometimes I felt as if I had known and loved him forever—indeed, the nine years of our off-and-on relationship totaled more than a third of my life. As his visit drew closer, I began to dream about him and awoke in the morning missing him. I couldn't wait to see him.

On the night of August 11, 1971, I trudged up the steep road to my home above the Berkeley campus. By the glow of the streetlight, I saw John on the porch going into my house. I ran the rest of the way up the street, full of joy and a little surprised he was back from the Yukon already. Breathless, I bounded inside. But John wasn't waiting in the living room. I ran down the stairs to my room and opened the door, expecting to see his bright blue eyes and smiling face. The room was empty.

Was John hiding to surprise me? Was this a joke? I searched every room and questioned my housemates. They hadn't seen him. He was not there. It was not a joke. Dee suggested that since I'd been anticipating his visit so eagerly, I must have imagined seeing him. There seemed no other possible explanation. Disappointed, I continued to write my thesis and wait for John's arrival.

A week later, as I was paging through the *San Francisco Chronicle* at breakfast, my eye was caught by a headline that read THE AGONY OF AN AVALANCHE. Next to the headline was a picture of John. I read that John Henry Hall, Susan Deery, and Stanley and Lucille Borgen Adamson had been buried in a huge avalanche on Mt. St. Elias.

No. No. They couldn't be dead. Feeling like I was about to faint, I called John's parents in Portland. After many rings, John's mother answered the phone with a weak hello. From her voice, I knew it was true. I felt sick and could barely manage a few sympathetic words to her before I began sobbing. After I hung up, my sobs turned to screams. I collapsed on the floor and screamed until my throat was raw. It wasn't true. It couldn't be true. How could John be gone? How could my four friends be erased by snow and ice? I couldn't stop crying until Dee came home for lunch and hugged me for a long time. My heart was broken, and I ached with memories of John.

John had taught me about the wilderness, the mountains, the world. And now all his brilliance and love were buried forever beneath tons of snow and ice. I was torn apart by irrational guilt that I had not been there and somehow saved John and my other friends, and by relief that I hadn't died, too.

The avalanche had buried my friends on the afternoon of August 11, only a few hours before I thought I saw John on the porch of my house. I tried to find solace in the idea that in some incomprehensible manner he had come to say good-bye, but there was little comfort there. I couldn't accept that John and I wouldn't share love or adventures or anything else ever again. Finally the tears ran out and I became numb. I continued doing all the things required of me; the memorial service in Portland, visits with the parents of my lost friends, and the writing of my thesis came and went as I wandered in a murky nightmare.

Toby Wheeler, the only survivor of the St. Elias avalanche, stopped by our house a month later. With closed eyes, I listened to Toby's story.

Everything had been going really well, he told me. They'd climbed Logan; they were feeling strong and happy. It was a hot, sunny afternoon, they were heading up a moderate snow couloir, and it seemed safe enough. All of a sudden they heard a loud crack and saw a cloud of snow far above them. The cloud got larger and was heading straight toward them. They tried to run but there was no time. The avalanche gathered speed and size and then they were in it, blocks of ice bigger than houses. Everything was dark and roaring and Toby was sure he was going to die. His pack was torn from his shoulders and he swam for the top like one is taught to do in an avalanche course. When the avalanche stopped, the snow and ice set up like cement in seconds. Miraculously, Toby was on the surface, alive. He moved his arms and legs. They seemed fine. He looked around. The sun was still shining, but everything was different. The glacier was eerily quiet. The rope at his waist connecting him to Stanley pointed straight down into the frozen snow. His four companions had disappeared. There was not a living thing in sight.

Toby told me he'd unclipped from the rope and dug at the hard ice until his fingers were bruised and bloody. As the sun dropped in the sky, the temperature plummeted and he was shivering, his only companion his long shadow stretching across the icy surface. There was no way he could dig anyone out alive. He had no food, water, fuel, or shelter. Everything, everyone, was buried.

Somehow Toby found the strength to walk out, and two days later he reached their camp on the main glacier. They'd left a radio there and Toby called for help. The bush pilot came in and flew him out just before a weeklong storm. There was no hope for the others. A physical pain coursed through my body as I listened to his story and once again felt the finality of the loss.

Toby told me he still wanted to do the Endless Winter, and asked if I did, too. I didn't know. My feelings about mountains had undergone a 180-degree change. Rather than thinking of the regions above timberline as benign havens, I now saw mountains as unfeeling adversaries that doled out, with random malevolence, the terror of avalanches and the ultimate punishment of death for bad luck or a small error in judgment or technique. Bereft and overwhelmed with anguish, I now feared high mountains and was not especially eager to spend the next year and a half climbing them.

On the other hand, the Endless Winter was planned, and however little enthusiasm I felt for it, I had even less interest in anything else. Maybe being with Toby would connect me to John and my friends, to all I had lost. Maybe a journey to faraway lands would distance me from my grief and from my mother's neediness. I hadn't yet learned that wherever I went, my ghosts would accompany me.

CHAPTER 12 ▲

The Isenberg Girls

Chicago 1959

A photo of four beautiful, dark-haired young women gathered around Grandma and Grandpa hangs in a dim corner of our dining room. After school one afternoon, I take the picture off the wall and bring it to the window to see it better. My mom, Gertrude, the oldest of the four Isenberg sisters, is standing on the left; then come Grandma, Grandpa, and Sylvia, the next oldest sister, on the right. Mom and Sylvia are wearing suits. The two younger sisters, Shirley and Ruth, sit in front, wearing flared floral dresses. The sisters look strong and proud, radiating happy girl energy.

I take the picture upstairs to our room.

"You all look so pretty," I say, showing it to my mom.

"Yes, we do, don't we?" She smiles at the photo.

"Why are you so dressed up?"

"My sisters were probably going out on dates. Maybe I had a concert. All the boys were after my sisters then," she says. "But I didn't know anything about men or love in those days. I just played my violin."

She takes the photo from me and looks at it more closely. "Sylvia, now, she had a good head on her shoulders. Always had the ideas and I followed her. Ruthie's the beauty. Did she ever have men chasing her! And Shirley, the baby; she married a fine man, too. My sisters are all smart, smarter than I am. They all chose husbands who take good care of them. But not me. I fell in love with love."

I hold my breath, hoping Mom will say something about my father. But she puts down the picture and picks up her book.

I study the photograph. It's true: my aunts' lives are so very different from my mom's. Each has two perfect children and a devoted husband. I'm jealous of my cousins' orderly lives. My aunts stay home keeping their houses clean, cooking delicious meals, and playing bridge.

My mother's family in the front yard of the Isenberg home in Davenport, Iowa, 1942. *(From left, standing)* My mother, Gertrude; my grandparents, Betty and Isadore; Aunt Sylvia. *(Sitting)* Aunt Shirley and Aunt Ruth.

Before she got married, Aunt Ruthie was my special friend. She took me along on her dates—to the beach, plays, and restaurants where her boyfriends let me order anything on the menu. Ruthie gave me new clothes and taught me how to keep them clean and be a proper young lady. When I was with her I could pretend I was like other kids.

But then she married Uncle Sy and had her own kids. I like Sy and my cousins, but now it's different between Ruthie and me. Like last weekend when I stayed at her house.

At breakfast, she gave my cousins bowls of cereal heaped with blueberries. I waited for her to give me berries, too, and when she didn't, I asked if I could have some.

"Blueberries are your little cousins' favorite fruit. They're expensive and hard to come by this time of year," she said, and handed me a banana.

The Endless Winter in Africa ▲

1971

Bound for Gonder, Ethiopia, in December of 1971, I boarded the Pan Am jet with an ice axe in my hand, climbing rope slung over my shoulder, and heavy mountaineering boots on my feet. I wore a stylish blue and white tailored dress, trying for a respectable look in case anyone questioned the gear adorning my body. Our budget didn't include excess baggage charges, so we'd carefully weighed out our forty-four-pound allowance and managed to wear all the rest. I'd declined the flight attendant's offer to hang my jacket in the plane's closet, not wanting her to discover the pitons, ice screws, and hammer in the pockets.

Sinking into my seat between Joel and Toby, I was worried we had forgotten something. During the next year and a half, we'd be traveling to some of the coldest and hottest places on the planet and we had tried to bring along the essentials. Joel, nearly blind without his eyeglasses, had even brought along three extra pairs.

My excitement at departing for our great adventure was tinged with exhaustion and sadness. All I had done for the last few months was work on my thesis, grieve for Grandpa and my friends, and eat junk food. For the thousandth time I wondered if I'd made the right decision in carrying on with the Endless Winter. But there was no going back now.

Flying from San Francisco to the dirt landing strip near Gonder, we traveled back centuries in time. A sign informed us that Haile Selassie, King of Kings, Conquering Lion of Judah, Emperor of Ethiopia, welcomed us to his country. Vultures languidly circled above the parched, stony ground. While we waited for our bags to be unloaded, we were surrounded by a gaggle of ragged urchins clamoring to shine our shoes, guide us around town, sell us Chiclets. I naively offered a small boy some mints from the plane, and word that we were giving out free candy spread quickly; before long I was mobbed by a throng of begging

Leaving San Francisco, bound for fifteen months of climbing around the world, and trying for a respectable look, I wore a stylish blue and white tailored dress. *(Photo by Ludwig Blum)*

children. Once my candy was gone, the kids pushed in close, pulling at my clothes and demanding more sweets. At the very beginning of our great adventure, I felt disconsolate.

From Gonder, Joel, Toby, and I took a bus to the dusty market village of Debarek. Our plan was to buy food, hire a guide and horses to carry our loads, and walk for a week to the Semyen Mountains, located on a high arid plateau known as the Roof of Africa. After doing some easier climbs, we would try for the summit of Ras Dashan, at 15,158 feet the highest point in Ethiopia.

Our first day's hike took us through cultivated fields containing so many rocks that the local farmers didn't even try to clear them out. Instead they planted *teff,* the smallest grain in the world and the basis of the Ethiopian diet, in the bits of soil scattered amid the stones. A strikingly beautiful woman with high cheekbones, dressed in tattered sackcloth, offered us some *tej,* the local honey wine. Her handsome face lit up when we thanked her with a few words of elementary Amharic.

The countryside became even more rugged as we made our way to the high plateau. One of Africa's most impressive escarpments girdles the northwest side of the Semyen, plunging down from the plateau for nearly a vertical mile. A herd of the rare black-bearded goat called Walia ibex grazed on the precipitous cliffs. Troops of gelada baboons, red hourglasses on their chests and capes of long hair flowing from their shoulders, grunted greetings and groomed one another. The rare Verreaux's eagle, a host of vultures, and other birds of prey glided over our heads.

The long, hard days of trekking the rough Semyen trails calmed my agitated mind. The nights, though, I dreaded. In my dreams huge avalanches thundered down on me, and I saw John Hall forever frozen in the ice. Sometimes I dreamt that I went along to St. Elias and saved everyone. And other nights I dreamt that I died, too. I awoke drained and despairing, the pain of my losses agonizingly present in the Ethiopian dawn.

After four days of trekking across the harsh, arid countryside, we made camp on soft grass between two clear streams flanked by giant lobelia bushes and ancient trees dripping with moss. We were in position to ascend the Endless Winter's first peak, Buahit, at 14,800 feet the second-highest point in Ethiopia.

Knowing I hiked more slowly than Joel and Toby, I got route-finding directions from Toby just before the two of them hiked quickly toward Buahit Pass. After they disappeared, I ascended the steep trail at a steady pace and thought about our team.

After three weeks of travel together, my Endless Winter family remained distant. I was getting to know Joel, and though he was cordial, we weren't close. Toby was aloof and sometimes even scornful. I wondered where his negativity was coming from and why it was directed at me.

On the arid plateau, I followed Toby's directions, only to be stopped by an impassable break in the escarpment. Worried, I retraced my steps and found an easy trail heading up to where Joel and Toby were having lunch. They had already summited Buahit and were on their way down.

I congratulated them and apologized for being late.

"What's wrong with you?" Toby asked disdainfully. "We waited forever before we went up."

"I took the route you pointed me to," I said, hurt by his tone. "Your directions didn't work."

"You didn't listen," he retorted. "The trail's obvious."

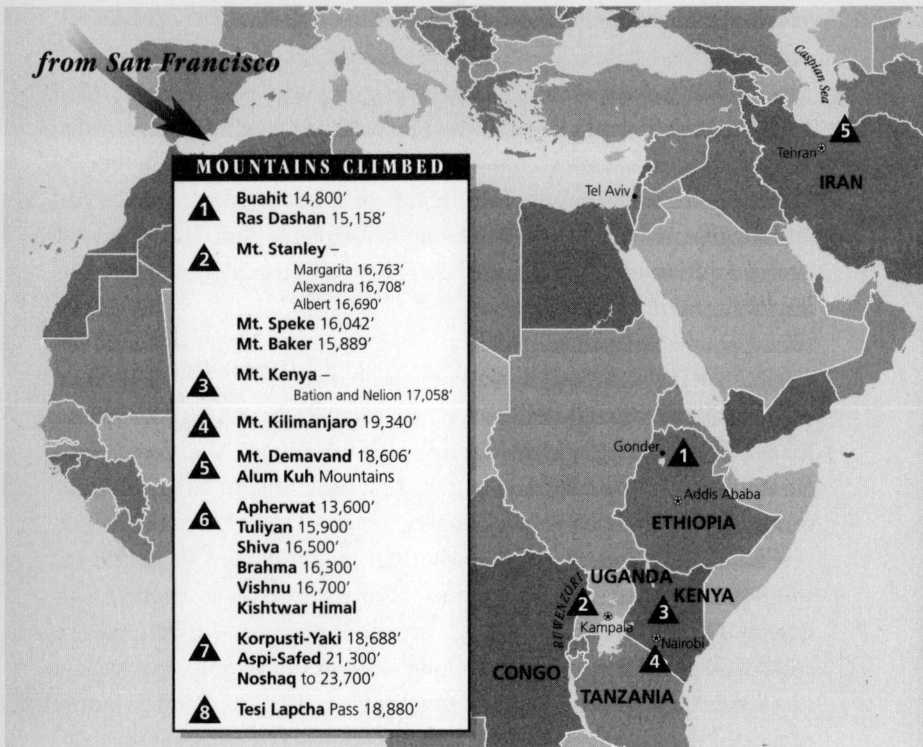

I started to remind Toby that he'd told me to go the wrong way, but I bit back my words, wanting to keep the peace. I listened eagerly as Joel and Toby told me about the fabulous views they'd had from the summit as we shared a lunch of Bulgarian jam and Danish cheese on English crackers with Malaysian pineapple and Israeli chocolate, all purchased from the Gonder market.

After we ate, I told them I was going to the summit and would meet them back at our camp. I started up.

"You need to come down with us, Arlene," Toby called after me. "There's not enough time for you to go to the top."

"It's early afternoon, the weather's perfect," I shouted down. "You don't have to wait for me."

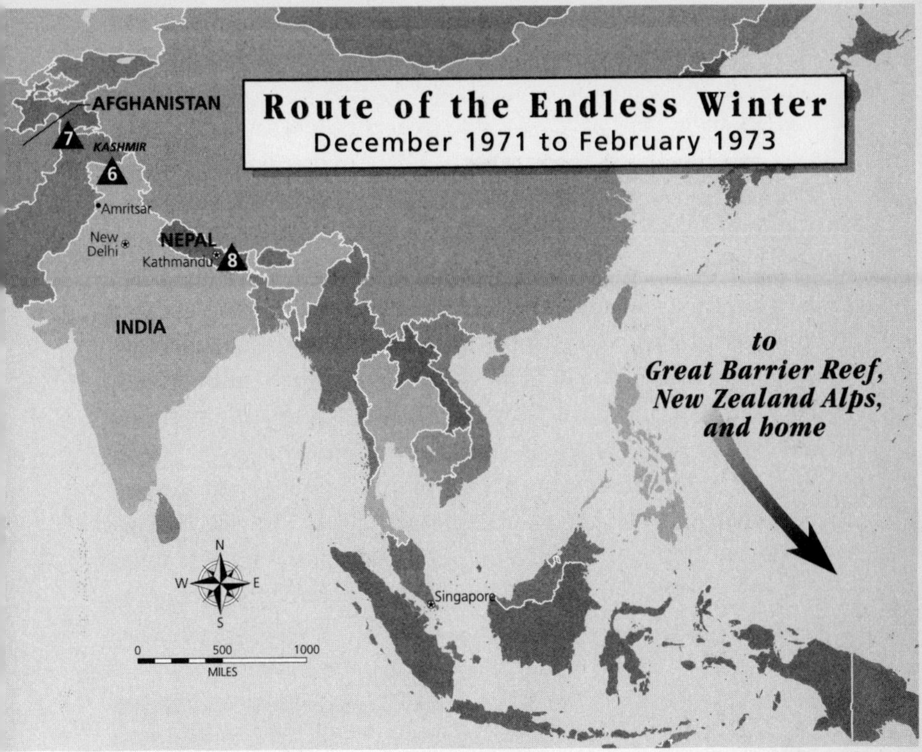

Route of the Endless Winter
December 1971 to February 1973

"You're moving too slowly," Toby insisted. "We're heading down and you're coming, too."

My patience snapped. "You can go wherever you want—to hell as far as I'm concerned," I yelled back at him. "But you can't tell me what to do. I'm going up the mountain."

I stomped off, my heart pounding wildly. I wasn't turning back without summiting. A burst of adrenaline propelled me to the top in twenty minutes. As I took out my tRNA flag for a photo on my first Endless Winter summit, my anger ebbed and was replaced by embarrassment at my outburst. In my family, people exploded when they were mad. I guessed, from Joel's and Toby's shocked faces when I strode off, that they weren't accustomed to being shouted at.

I headed back down, rehearsing my apology. To my surprise, Joel and Toby were still at the shoulder where we'd had lunch.

"I'm *really* sorry," all three of us said at once.

"We thought you'd never speak to us again," Joel said.

"You sure made good time," Toby said. "Congratulations."

Pleased that everyone was still cordial after our disagreement, I chalked the incident up to the fact that we were just getting to know one another. We trooped companionably down four thousand feet in a mile to our camp by a medieval church.

Apparently, few *ferengi,* or foreigners, visited this region. As we sat at our campfire, a crowd of local people crouched nearby, staring at us. We felt uncomfortable until Toby motioned them to join us for tea. Then a herder offered to sell us a goat for our dinner.

At dark, most of our visitors went home, beaming and waving, but two young men stayed behind to clean and roast the goat. We invited them to share our meal. With our guide translating, I pointed up at the huge full moon and told them that men from our country had walked up there.

Villagers watch us begin to set up our camp, 1971. Ras Dashan, Ethiopia's highest peak and our objective, is in the upper right.

"To walk on the moon." The young man shook his head, bemused. "That cannot be possible."

The next day we ascended more than six thousand feet and camped out at dusk on the rocky summit of Ras Dashan. I slept soundly and dreamlessly all night, snuggled between rough boulders. Awakening before dawn, I brushed ice crystals from my sleeping bag and felt a lightening of my spirits as the morning sun illuminated the glorious world around me. The panorama of the high Semyen peaks blending into the mists of the lowlands was so immense and magnificent that my pain seemed small in comparison. I felt more at peace than I had since Grandpa's death. It was good to be back in a high place.

In early January 1972, after a flight to Kampala, Uganda, and a train ride, we arrived in lush tropical Fort Portal, the gateway to the Ruwenzori, the legendary Mountains of the Moon.

In the second century, the Greek geographer Ptolemy postulated that snowcapped mountains on the equator in Africa were the source of the life-giving waters of the Nile. During the exploration of Africa in the nineteenth century, the "discovery" of these fabled mountains became one of the most sought-after explorer's prizes on the continent.

The Ruwenzori range is like an arctic island rising directly from the equator, and the warm air meeting the cold heights creates perpetual mist. This thick curtain prevented the Ruwenzori from being seen by Europeans until Henry Stanley, who later found Dr. David Livingstone, sighted them in 1888 during a brief clearing of the clouds. Numerous explorers were turned back by the trackless rain forest and thick fog, until 1906 when an Italian expedition led by the Duke of the Abruzzi succeeded in climbing the three major massifs that dominate the range. These peaks—Mt. Stanley (16,763 feet), Mt. Speke (16,042 feet), and Mt. Baker (15,889 feet)—whose enormous glaciers feed the Nile—lie just where Ptolemy placed the Mountains of the Moon on his ancient maps.

As we got ready for the trip in Fort Portal, my relations with Toby continued to deteriorate. For reasons I didn't understand, he objected to my suggestions and made sarcastic jokes at my expense. I avoided him whenever possible and fell into a comfortable alliance with Joel, with whom I worked well. Caught in the middle, Joel diplomatically managed to maintain good relations with both Toby and me.

Shopping for the long list of items suggested by the Mountain Club of Uganda, Joel and I were able to find everything we needed except the required "quarter pound of ground nuts a day" per porter. We scoured the markets for packages of chopped nuts, confused as to why the porters needed them. Finally a shopkeeper told us that ground nuts are nuts that come from the ground—and are also known as peanuts, piles of which were everywhere in the bazaar.

Although the local people were friendly and helpful, control of Uganda had recently been seized by Idi Amin, a ruthless dictator. The main street of Fort Portal had been reduced to gravel from the weight of Amin's tanks, and as we walked into town every day, we passed the same two dead bodies lying in a creek. Apparently no one dared move them.

Leaving Amin's gruesome relics behind, we headed toward the Ruwenzori with five porters from the Bakonjo tribe. To my relief, Toby usually walked ahead with the porters, and Joel and I followed. Each day, we encountered mist, rain, sleet, and snow, all rapidly alternating with warm, bright sunlight. Sometimes we hiked on thick carpets of soft yellow, green, and orange mosses dotted with delicate wildflowers, amid forests of strange and lovely giant lobelia and groundsel. The intense ultraviolet radiation at high altitude on the equator is said to cause small plants to mutate and grow to the size of trees. More often we followed a patchwork of tussocks of dry grass surrounded by deep mud.

Indeed, the trails demonstrated the splendid diversity of mud: thick, slippery red mud; squishy, sticky black mud; gooey, slimy brown mud; honest mud that just dares you to step in; and waist-deep mud covered by a deceiving layer of brilliant moss. The bogs were similarly varied: horizontal bogs, inclined bogs, vertical bogs, and even an amazing overhanging bog—a steeper-than-vertical rocky area covered with mud, lichen, and dripping moss held together with tree roots.

When we undertook the crossing of the notorious Bigo Bog, I was cautious at first, carefully leaping from tussock to tussock. Soon I slipped ankle-deep into the mud. Giving up all hope of keeping my boots dry, I strode boldly down the center of the trail, certain that I could get no muddier. A moment later, I sank to my knees in the oozy mire. With difficulty Joel helped me struggle out. We arrived at the next hut, coated with drying mud, to find our porters waiting, completely dry and free of dirt. While we foreigners wallowed, the Bakonjo had the mysterious ability to tred on the surface of the bog.

Mt. Stanley, the highest peak in Uganda, is a complex massif with three icy summits. Margarita at 16,763 feet is the main summit; Alexandra and Albert are slightly lower. On a rare sunny morning, we set out to climb Margarita. We were making good progress up a slope of ice-coated rocks when a wild hailstorm assaulted us. When lightning began to strike nearby, we retreated. Back at the cold, damp hut, we climbed into our sleeping bags to wait out the miserable weather. I managed to read all of *The French Lieutenant's Woman* before emerging from the bag for a very necessary trip to the loo. Then I read aloud to Joel and Toby from the logbook signed by past visitors to the hut:

"The whole track is a bog except where it's a quagmire."

"I have been to the Ruwenzori five times and never seen any of the peaks."

"It's like the trails have signs saying 'Don't walk on me.' If you do, the punishment is imprisonment in waist-deep mud."

The next day, we again tried to ascend Margarita. Hail pelted down as we climbed cautiously up the six-hundred-foot unstable rock slope. But this time there was no lightning. The fog, eddying around us, was so thick that I could barely see my own feet. We roped up and ascended five hundred feet of slippery rock, chipping ice off many of the holds. Our reward was a gentle snow slope, which we followed to a narrow summit ridge.

Heavily glaciated in 1972, Mt. Alexandra (16,708 feet), on the border of Congo and Uganda, is the second-highest peak in the Ruwenzori.

Breathing hard, I reached the ridge of Margarita a little behind Joel and Toby. Magnificent peaks shimmered through occasional breaks in the fog, their fantastic ice-feather cornices leaning out many feet over empty space.

I knew a cornice can form when the wind always blows snow in the same direction across a high mountain ridge. The snow builds up and hangs out in a formation that looks like a mountaintop, but has only air beneath it.

Taking care to avoid the danger of stepping through the cornice on the summit ridge of Margarita, I slowly made my way to the highest point. Once on the very top, I took out my tRNA flag for a photo. In spite of my promise to Nacho to continue to think about my research, I felt very far from it.

From Margarita, we crossed the unmarked border and climbed Albert, the third-highest summit of Mt. Stanley, and at 16,690 feet the highest point in the neighboring Congo. Up here, Uganda and

Congo were indistinguishable lands of snow and rock, not what one would expect for equatorial African countries. On the way down Toby and Joel climbed Alexandra at 16,708 feet, the second-highest peak of Mt. Stanley. I was content to stay below and take dramatic photos of the two of them on the summit.* Two rappels brought us back to our hut.

Having heard that the next day's trail to the Mt. Baker hut was lengthy and rough, I got up early and left ahead of the others. For two hours I slogged through deep mud and swung down steep slopes on branches, roots, and vines. When the trail improved so it didn't require as much of my attention, I started to mull over the tension between Toby and me. What was the problem? Did Toby scorn me for my low spirits? Or might it be that I reminded him of his ordeal on St. Elias? Could he feel guilty about surviving? Maybe if we discussed the avalanche, we could clear the air and our relationship would improve. I'd promised John Hall's mother that I'd share with her what Toby told me about the Logan trip, but so far he hadn't said a word. Perhaps, I thought, he was waiting for me to bring it up.

Just then Joel and Toby joined me, and without thinking, I blurted out, "Toby, don't you think John Hall would have loved the Ruwenzori? I wish he were here with us in this outrageous place."

Toby said nothing.

"I hope it isn't painful for me to bring this up, but I'd appreciate hearing a little more about your Logan or St. Elias climbs," I continued, feeling increasingly awkward.

Still not a word from Toby. It seemed like everything had fallen silent—the birds had stopped singing, the insects buzzing, and the trees rustling. I felt compelled to fill the enormous silence.

---

*In 2007, thirty-five years after our ascents, the magnificent glaciers of the Ruwenzori mountains are much diminished, a victim of global climate change and the CO_2 being added to our atmosphere by our increased use of energy.

Current predictions suggest that the equatorial glaciers will disappear within the next two decades unless we find a way to decrease our use of energy and reverse this trend.

"I know it's difficult for you, but it would mean so much to me to know what you remember about the days before the avalanche . . . anything at all . . . please."

Now that I'd finally started saying what I had been thinking for so long, I couldn't seem to stop, even in the face of Toby's obvious discomfort.

"All that is the past," Toby said flatly. "It is over and done with. I don't want to think about it anymore. I live in the present, not the past. I look forward, not backward."

Toby turned to Joel. "So, what do you think about trying a new route up the face of Mt. Baker?"

Feeling hurt and invisible, I barely listened to their animated conversation. My heart ached. Here I was halfway around the world with the person most closely connected to my friends' deaths and he refused to talk about them. Did I have any right to expect Toby to go back to his terrible experience? Maybe not, but I did have a right to expect him to be civil to me. Whatever was going on with Toby certainly wasn't pleasant for me or good for our trip and I had no idea what to do about it.

After lunch, I set off alone again through the Ruwenzori forest. Part of the trail was a tunnel beneath a dark canopy of dripping vegetation. By the time I reached the Baker hut, I was thoroughly despondent and suffering from a head cold and the beginning of trench foot, a painful condition caused by continual cold, wet feet.

In spite of the tension between Toby and me, we'd come all this way to go climbing and that's what we did. The next morning, Joel and Toby went off to attempt the second ascent of the northwest face of Baker, a route that was too technically difficult for me. They later reported climbing several hundred feet of extremely steep, brittle ice to reach the summit. Joel, who'd previously climbed the notorious Black Ice Couloir on the Grand Teton, said it was the most difficult climb of his life.

The next day, I headed up a relatively easy trail along the south ridge of Mt. Baker by myself. The crux was a thousand feet of heavily overgrown, seemingly vertical mud, referred to euphemistically in the guidebook as the "trail." I focused on finding my way without getting slathered in mud.

My mood improved as I moved steadily up slabs of rock and gentle snowfields. At last I was on top of Mt. Baker, looking back at the enormous ice cornices overhanging the Stanley plateau. Below me, a huge bank of clouds rolled in from Congo. I flew my tRNA flag and felt some pride at having found my way up here, to nearly 16,000 feet, on my own. Descending the slippery rock in alternating snow and sun, I resolved to try to forget the pain of the past as Toby suggested. I would do my best to get along with him and to enjoy our wonderful present adventures.

A week later the three of us were back in Fort Portal, relaxing in the comfortable home of the Heersinks, new Canadian friends we'd met at the first hut who were volunteering at the local hospital. The night before our return to Kampala, I was reading a book on the terrace after the Heersinks had gone to bed when Toby walked in.

"The Endless Winter is over," he abruptly announced.

"What are you talking about?" I was stunned.

"The three of us are incompatible," Toby continued, looking at the floor. "Joel and I are going to try some hard routes on Mt. Kenya and Kilimanjaro on our own."

"You can't just leave," I protested. "You can't do the trip without me."

"You're too slow," Toby said. "We're taking the early train to Kampala tomorrow. You can do what you want." He turned and walked away.

Holding back my tears, I ran outside and plunged into the bush. When I was far enough from the house so I couldn't be heard, I sat down on the ground and cried. My world was annihilated. Joel and Toby were leaving. Grandpa, John, and my other friends were dead. I felt as abandoned as if I were the last person left on earth. Then through my tears, I felt something warm next to me. It was the Heersinks' tabby cat. I stroked her soft fur and the cat began to purr. The rhythmic sound calmed me. I opened my eyes and saw the myriad stars of the African night. I realized I was not an abandoned child but a competent twenty-six-year-old woman. I had a PhD from Berkeley. I'd helped get Grace down from Denali. My life wasn't over. Maybe I could get a job teaching at the University of Uganda in Kampala. Although the Heersinks were new friends, maybe they would have an idea of how I could stay in Africa on my own.

Drained, I crept back into the house and fell into an exhausted sleep. The next morning I awoke and heard Joel and Toby leave for the train to Kampala, but couldn't manage to get out of bed until lunchtime.

Taking a look at my pale face, Sue Heersink gave me a sympathetic hug and a glass of Waragi, the local gin, and passion fruit juice. Sitting on the terrace above their lush garden, watching the birds drink nectar from the passionflowers, I sipped my bittersweet drink and told Sue and Ben what had happened. They encouraged me to go to Kampala and try to persuade Toby and Joel to reconsider. The thought of seeing Toby made my stomach lurch, but I decided it was worth trying to talk to Joel.

After taking the train to Kampala, I found Joel alone and nervously invited him to tea at Bat Valley, a popular Indian restaurant. As we sat at an outdoor table eating Indian sweets, the huge trees towering above us exploded with tens of thousands of bats taking wing for their nocturnal foraging. Endless streams of dark shapes continued to emerge from the mammoth trees as we talked. Joel listened as I made my case: I thought he and I climbed well together and enjoyed each other's company. The trip was originally my idea; I had set up our itinerary and made most of our contacts with the local climbing clubs. I didn't feel safe traveling alone in Africa and I couldn't climb the peaks on my own. I stopped, held my breath, and waited for Joel's response.

He thought for a minute and then smiled. He told me he understood my not wanting to be left on my own and that he'd thought of this as a temporary separation, not the end of the trip. "I'll try to work something out with Toby," Joel said. "If not, I'll climb with you. Toby's an independent fellow."

Joel's decision filled me with relief and happiness. Our friendship was restored and I wouldn't have to deal with Toby for a while. I felt hungry for the first time in days and we ordered a huge dinner of delectable Indian curries. Joel later shared a letter he'd written the next day about our situation:

> It's funny how on the surface everything is perfect: the climbing, meeting people, doing things, while underneath we have been at odds with one another. We had zero communication until last night. We have all learned a good lesson on how not to live with people; remaining silent while being mistreated until an explosion occurs is a poor way we all knew, but no one opened up. Arlene and I have finally established some rapport and understanding.

Joel told Toby of the new plan, and Toby said he would buy a Land Rover and explore the game parks of East Africa. So Toby traveled on his own while Joel and I swung leads on a technical climb up Nelion and Bation, the two highest summits of Mt. Kenya (17,058 feet), and hiked all the way around the Kenya massif. Our final ascent was up 19,340-foot Kilimanjaro in Tanzania, where I briefly sunbathed in a bikini on the very top of Africa.

Joel and I then headed to the romantic island of Lamu off the Kenya coast for a break. Sharing adventures and living so closely day after day, I developed a crush on Joel, an attraction that wasn't as requited as I might have hoped. Still, we climbed and traveled together easily as good friends.

As Joel had written, to all appearances the Endless Winter was going extremely well. We'd ascended the highest peaks in Ethiopia, Uganda, Congo, Kenya, and Tanzania with only the gear we could bring on an airplane. But the real story of the trip was more ambiguous. Embarking on a fifteen-month expedition with the only survivor of an avalanche that had killed my first love and three other good friends was, to put it mildly, probably not a good idea. But I wasn't ready to go home.

After a week in Israel, Joel and I unexpectedly met up with Toby and Dave Graber, our fourth team member, at the Lod airport in Tel Aviv. Taking our accidental meeting as an omen that all would now be well among us, we boarded the plane together for Iran and the high mountains of Asia.

Joel and I relax in the saddle between Kilimanjaro and Mawenzi, Tanzania, 1972.

Chapter 13 ▲

Learning to Swim
Chicago 1955

School's been out for two days and I feel like I'm trapped in a steam bath. Grandpa tells me the temperature is 99 degrees and the humidity 99 percent. I lie in the shade of our cherry tree, listlessly reading a pile of Nancy Drew mysteries. Rivulets of sweat roll down my sides, meeting the ants crawling up. There's no escape from my sticky body, the muggy air, and the TV noise blaring out the open window. I've got to do something different this summer or I'll die of heat and boredom. I go inside where Grandma is lying on her couch.

"Grandma, can I take a class at the YMCA this summer?" I ask. "Maybe I could learn to swim."

My grandmother's gaze moves from As the World Turns *to me. "Public pools are dangerous. You could get a disease."*

"Sherry swims at the Y. She's not getting any diseases."

"Polio, flu, you never know," she says. "And what does a Jewish girl need with the Young Men's Christian Association?"

"Then can I go to the store to buy an ice-cream cone?" I beg. "Please."

"You know I don't want you crossing 100th Street. There are nutcases behind the wheel out there. A car could hit you. You should stay home."

I sigh out loud, retreat to the shade of the tree, and pick up my book.

But I am determined to learn how to swim. First I persuade my mother that swimming would be good for my health. Then I convince my grandfather that it would help me make new friends. And finally, in the middle of a severe July heat wave, my grandmother gives in, probably just to stop my nagging.

Mom takes me to the Y to sign up. While she fills out the forms, I study a chart that shows the requirements for the various swimming levels, from Jellyfish to Shark. I dream of becoming a Shark.

The first day of class I nervously lower myself into the heavily chlorinated indoor pool. A tall ten-year-old, I tower over the other Jellyfish, most of whom are five or six. When I lift my feet off the bottom to try to float, I sink. I'm drowning, *I think as water fills my nose and mouth. I burst out of the water, a sputtering giant, frightening the little girls in my class.*

I go to the Y every day. Two weeks later I can hold my breath underwater and float. I'm thrilled when the teacher says I'm ready to be a Guppy.

Then I learn to tread water and become a Tadpole. Taking my first crawl strokes in this new class is scary, but the teacher supports me, patiently showing me how to move my arms and breathe. I practice every day for a month and I can pass the tests for a Minnow. Swimming after school for three months, I learn all the strokes well enough to become a Fish. I learn water ballet, which I love, and how to dive off the high board. I'm a Flying Fish. I join the swim team and practice for hours every day.

Just before my twelfth birthday, my dream comes true: I am awarded the long-coveted patch and certificate that proclaims to the world that I'm a Shark. I can't believe it's me.

The Queen of Tenacity ▲

1972

AS WE FLEW INTO TEHRAN at dawn, the symmetrical summit of Demavand, at 18,606 feet the highest point in Iran, shimmered amid frothy violet clouds. Dozens of other peaks poked through the mist in the distance. Most of the Iranian mountains had rarely been climbed; some not at all. Maps were few and information about past explorations scanty. Anything was possible during the months ahead.

Downtown Tehran reminded me of Chicago—urban, loud, almost frenetic. We struggled through the crowded streets looking for climbing supplies, but the things we needed, like kerosene and alcohol for our stoves, and sugar, tea, and oatmeal for our breakfasts, were hard to find amid the stalls of French perfume and imitation Dior handbags.

An Iranian who spoke fluent English attached himself to our group, offering to help us find the supplies we needed and the bus station. On the way to the market, we just happened to pass the store of his brother-in-law, who just happened to deal in fine Iranian carpets. After many mugs of local beer, Toby, who came from a wealthy family and apparently had an unlimited budget, handed over $2,000 for "bargain" carpets. This seemed an extravagant purchase to the rest of us, who were living on $200 a month. Our "guide" and his brother-in-law pocketed the money and vanished, leaving us in the bazaar with Toby's carpets, but no kerosene or bus tickets.

As we made our way through the marketplace, I noticed that Iranian men didn't seem able to walk straight—they kept bumping into me and their hands ended up in inappropriate places. To reduce this harassment, I took to walking between Joel and Dave, both of whom were tall and could run interference. But I continued to get pinched by men on the street until I realized that the short comfortable skirt I was wearing, one that had attracted no special notice in Africa, sent the wrong signal in a Muslim country. Once I began wearing loose pants or a long skirt, the unwanted attentions stopped.

After shocking the young man who worked at our hotel by saying we were not married, Joel and I decided it would be prudent to let people assume we were husband and wife. That, of course, led to the inevitable question: "Do you have issue?" It took us a while to realize they were asking about children, not a copy of *Newsweek.*

A few days after our arrival, we enjoyed a multicourse dinner at an inexpensive Russian café. Joel's stomach rejected the meal as we were leaving the restaurant and he sank to the curb, retching his dinner into the gutter. Moments later, I felt just as nauseated, but couldn't manage to vomit. Bent over in agony, I stumbled to our hotel and fell down on my bed. Every few minutes I staggered down the hall to the filthy, squat toilet, a hole in the floor with a raised footprint on either side. Between these desperate forays I couldn't even sip water. After two days of agony during which I sometimes felt death might be preferable, the acute food poisoning departed as quickly as it had come, leaving me weak but eager for the clean air and peace of the high mountains.

Meanwhile, Dave and Toby had not only managed to buy our supplies and bus tickets, but had also located the office of the Mountaineering Federation of Iran. The Iranian climbers graciously lent us skis for our Demavand ascent and suggested that we stay in their club hut at the base of the mountain.

A wild and bumpy ride brought us to the village of Raineh at 6,500 feet, from which we hoped to begin our climb of Demavand. In the center of this poor village—a scattering of mud huts without running water or electricity—a two-story wood-and-stone chalet stood behind a blue metal fence. This, apparently, was the Mountaineering Federation hut.

Entering the high-ceilinged living room, we were surprised to find ornate French furniture and lush Persian carpets. The rest of the chalet consisted of a modern kitchen, three bedrooms with queen-size beds piled with thick comforters, and a spotless bathroom with a choice of Western flush or Eastern squat toilet. We felt like royalty as we sat cross-legged on a precious rug, leaned against plump cushions, and devoured fresh, hot bread baked for us by the hut guardian. The Shah of Iran looked sternly down on us from his photo on the wall. The shah's stated goal of modernizing his country seemed to include lavish hospitality to foreign mountaineers.

Stepping out on the porch after dinner, I looked out through the blue metal fence separating our opulent dwelling from the drab village. From outside the fence, barefoot men and children dressed in tattered clothes stared at me with hunger in their eyes. Feeling painfully uncomfortable with the inequity between our circumstances, but not knowing what to do about it, I retreated into the chalet.

Our first day, we needed to gain 7,000 feet to reach the Demavand shelter at nearly 14,000 feet. Unfortunately, the horses carrying our gear sank chest-deep into the snow and had to turn back about halfway to the hut. We had no choice but to shoulder the loads ourselves and posthole through the deep snow. Keeping our energy up by munching continuously on pistachios—by far the best bargain in the Tehran bazaar—we marked our trail to the hut with nutshells.

I reached the hut just at dusk, a little ahead of Dave, but well behind Joel and Toby. As the first people to reach the hut that year, we had to dig out piles of snow that had drifted in during the winter before we could find a place to sleep.

The following day, facing a climb of 5,000 feet, we moved slowly in the intense cold and high wind. Clouds swirled in, obscuring our view. After climbing through deep snow for many hours we were still 1,000 feet below the summit. Dave, attempting his first mountain of the trip, had a bad headache and felt short of breath, symptoms of altitude sickness. Having just recovered from food poisoning, I didn't feel so great myself and offered to go back down with Dave. I was disappointed not to have reached the top, but pleased to have a new companion with a wry sense of humor and a sedate pace. Joel and Toby persevered on to the summit of Demavand and reported that they had to fight through 50-mile-per-hour winds to a high point reeking with the rotten-egg smell of fumaroles.

From Demavand we headed to the "Valley of the Assassins" near the Caspian Sea to attempt Alum Kuh (15,817 feet). Unseasonable spring snowstorms stopped us at 14,900 feet, burying the mountains and our hopes for further ascents in Iran.

Eating dinner in a hut in the Alum Kuh range. Iran, 1972. *(From left)* Joel, me, and Toby. *(Photo by Dave Graber)*

Before leaving Iran, we stopped at the Mountaineering Federation building in Tehran to thank our hosts. A friendly official showed us rooms full of new ropes, skis, and state-of-the-art climbing gear that the shah had donated to the federation. I thought of the poverty we'd seen all over Iran and, especially, the hungry eyes of the villagers in Raineh as they watched us through the blue metal fence.*

In the middle of May we flew from Iran to India's Vale of Kashmir—a verdant valley of lakes and flowers surrounded by snowy peaks. For only three dollars a day, we lived the luxurious life of Victorian aristocrats on holiday, staying in an intricately carved wooden houseboat anchored on the gently flowing River Jhelum. The British had constructed these fanciful, floating palaces to escape the summer heat of the Indian plains when the maharaja wouldn't let them buy land to build real houses.

Each morning a waiter clad in a formal white uniform ushered us to an elegantly set table on the boat's veranda. We enjoyed our five-course breakfast, entertained by flocks of quacking ducks, swooping hawks, and chattering women in colorful tunics. Children on their way to school, families on shopping trips, and an assortment of local merchants floated by.

We'd heard that the houseboat owners, whose customers had been British tourists for generations, profited more from kickbacks on what their visitors bought than from the modest room charges. Indeed, our tranquillity was often interrupted when vendors boarded our boat uninvited to offer us "bargains at below-dealer price." Anything we might want to buy—and lots we didn't—appeared on our deck, the merchants chanting, "You want Coca-Cola, mangoes, beer, embroidered shawls, rugs?" And failing that, "Hashish?"

---

*I later learned that in 1953 a coup initiated by the CIA had deposed Mohammed Mossadegh, the democratically elected leader of Iran, and put the despotic shah in power. Ironically, Mossadegh had been named *Time* magazine's 1951 Man of the Year; following the coup, he spent the rest of his life in jail or under house arrest. After installing the shah, the United States supported him in suppressing democracy in Iran, thereby setting the stage for the fundamentalist Islamic revolution in 1979.

In spite of the serenity of the Srinagar landscape, Kashmir was still shadowed by conflict arising from the 1947 partition of British India into India and Pakistan.*

We had arrived in Kashmir in spring 1972, a few months after the 1971 war between India and Pakistan over Bangladesh. At a local mountaineering club meeting, the Kashmiri climbers readily expressed their bitter feelings about Partition. We were dumbfounded when Mr. Mallik, a spectacled, mild-mannered Kashmiri climber, told us he had spent five years in solitary confinement in an Indian jail for trying to hijack an Indian helicopter to Pakistan during a war in 1965.

"But you're here to climb, not listen to our troubles," said Mohammed Ashraf, the secretary of the club. He then invited us on a weekend ascent of 13,600-foot Apherwat, a snow peak above India's highest ski area at Gulmarg.

Two days later the bus dropped the four of us, along with Ashraf and Mr. Mallik, across from a sign reading HIGH ALTITUDE WARFARE SCHOOL.

Starting our ascent at dawn, we climbed for more than six hours to reach a shoulder of Apherwat, from which we could see the true summit just beyond. Ashraf looked at the summit through our binoculars and suggested we stop to eat lunch and enjoy the view of Nanga Parbat, the first 8,000-meter peak I had ever seen, showing above the clouds far to the northwest.

"Why stop?" I asked, surprised. "The top is so close."

"It's better to stop here," Ashraf politely advised. "You can eat your lunch."

When I repeated that we'd rather go to the summit than eat lunch, he shrugged and motioned for us to go ahead without him.

---

*Lord Mountbatten, who was in charge of dividing the territory at Partition, told Hari Singh, the Hindu maharaja of Kashmir, a predominantly Muslim state, that he must decide whether to join Muslim Pakistan or secular India. The maharaja declined to choose, saying he wanted to remain independent. Mountbatten believed this to be a big mistake.

"India and Pakistan will have their daggers drawn against each other forever over Kashmir," Mountbatten said, which has proved to be only too true. A series of wars over the disputed territory continue to devastate this once beautiful land. And now, in addition to their daggers, the two countries have their nuclear bombs poised.

Reaching the top a few minutes later, we saw Pakistani troops in stone bunkers on the far side of the peak, their machine guns pointed in our direction. Without further ado, we put our hands over our heads, turned around, and ran down the summit ridge. Machine guns were not an acceptable mountaineering challenge.

"The summit ridge of Apherwat is the Indian-Pakistan cease-fire line since the last war, a no-man's-land," Ashraf explained calmly when we got back to where he was waiting for us. He went on to tell us that if the Pakistani border troops had seen us, they might have considered us invaders. "I've heard they have orders to shoot on sight," he said.

"Why didn't you tell us that before?" I asked, still breathless from my dash down the ridge.

"I didn't want to disappoint you," he said.

Evidently, I had been so eager to get to the top that Ashraf didn't want to upset me with the bad news that the summit was not safe. This was when I began to realize that Indians, like many others in Asia, do not like to say no directly. Negative information tends to be conveyed indirectly, and one needs to look and listen carefully.

Having no intention of invading Pakistan, we proceeded to glissade down thousands of feet. On the way down, Dave collected lizards to send to a herpetologist at UC Berkeley; he later learned he had found three previously unrecorded species.

Back at the bottom, the Kashmiris apologized for the armed troops on Apherwat. They suggested we join them to try the first ascent of Tuliyan, a challenging 16,000-foot peak that was far from the cease-fire line. They had attempted it several times without success and hoped that our combined strength would get us to the top. I was thrilled at this chance for a first ascent. The Kashmiris said they would meet us in Phalgam to attempt the climb the next Saturday.

Leaving for Phalgam two days early, we found the bus station jammed with people speaking myriad languages at top volume.

"We are in a Tower of Babble," Joel quipped as we joined a crush of people trying to buy tickets.

"What time is the first bus in the morning?" I asked when we finally reached the ticket counter.

"Anytime," replied a magenta-turbanned Sikh.

"Does the first bus leave in the morning or the afternoon?"

"Yes."

"What time does the first bus leave?"

"It could leave at 7:00 A.M."

"Are there other buses?"

"At 10:00 A.M. and 12:00 and 4:00 P.M."

"When is the last bus?"

"Probably at 2:00 P.M."

"How many buses are there each day?"

"One or two."

"We want to catch the 7:00 A.M. bus tomorrow. When can we buy tickets?"

"At 8:00 A.M. when the ticket office opens."

Catching on at last that time is fluid in India, we got up at a comfortable hour, had a leisurely breakfast, bought our bus tickets at 10:00 A.M., and caught the first bus out.

Our small hotel in Phalgam recycled trash in a most original manner: Everything was thrown off the balcony and devoured by roving dogs and cows. Indeed, one wandering cow waited outside our room each morning for its breakfast—our morning newspaper.

Our Kashmiri hosts generously brought an amazing amount of food for this four-day eight-person trip: forty-four pounds of rice, nine pounds of ghee (clarified butter), twenty pounds of potatoes, fifteen of onions, five of salt, five of mutton, a huge bag of hot peppers, and eight small candy bars for the summit. I smiled when I tried to imagine eating a pound of rice, half a pound of potatoes, and my share of hot peppers every day.

After a leisurely breakfast, our caravan of five horses, three horsemen, four Kashmiri, and four American climbers set off. We walked for several minutes and then waited a few more as the horses grazed—again and again. For lunch, the Kashmiris made delicious chapatis from scratch—an undertaking of several hours. The fresh, hot flatbread was delectable, much better than the stale, packaged crackers we usually ate. I tried to still my irritation, telling myself not to be such an impatient Westerner. It really didn't matter when we arrived at our High Camp. After two days, we reached the lush wildflower-filled Tuliyan Meadow at 10,500 feet. According to Dave's diary for the day: "7am to 11am egg & puree breakfast; 11–12:30pm hike to the lake (our high camp); dinner from 3pm–10pm."

I tried to reconcile myself to our companions' leisurely style, but I knew that an early start was critical for success on a high, icy peak in a tropical latitude. Thinking we'd never make it if the Kashmiris made chapatis, I volunteered to cook and serve breakfast on summit morning.

Awakening at 2:00 A.M., I tried to remain in my warm sleeping bag as long as possible. By the light of my headlamp, I carefully poured alcohol into the stove, pumped, and lit it. The flame shot upward toward the tent roof. I pumped frantically until the flame diminished and the stove gave off a satisfying roar. As I made breakfast in the dark, I remembered the Denali flyer saying that women could come to Base Camp and help with the cooking. Now here I was, cooking, and climbing as well.

I managed to prepare hot drinks and porridge for all eight of us without burning down the tent or spilling the pot of boiling water. At 3:00 A.M. I tried to wake the others. At first my offers of hot drinks solicited only groans from the sleepy climbers, but eventually everyone got up and ate breakfast.

At a quiet moment just before dawn, Joel and I headed out first through the Tuliyan glacier. Following his steady, even steps, I fell into a hypnotic state, almost sleeping my way up the mountain. Then the sun rose, revealing a wonderland of ice and snow sculptures. Our crampons cut firmly and satisfyingly into the slope. I was happy to be climbing with competent and witty Joel. Our paces were now compatible: I thought Joel went slower when we roped together, and he claimed I went faster.

Traversing around a gendarme (a huge rock tower), I followed Joel up a foot-wide ridge with 70-degree slopes falling away on both sides. The ridge was so steep I often could put my hands on the slope in front of me for balance. But sometimes we came to a horizontal part, where there was nothing to hold on to. Placing each foot very carefully, I balanced along this icy tightrope, my heart pounding.

A cornice, extending several feet out over our heads from the ridge above, stopped us just short of the summit. Joel hacked away at the cornice for an hour as I belayed him from below, ducking pieces of flying ice. Finally Joel climbed up a six-foot vertical wall and through the hole he had cut in the cornice.

At Tuliyan Base Camp. *(From left)* Joel, Mohammed Ashraf, Nazeer Ahmad, Vijay Kumar, Mohammed Aslan, and Dave. Kashmir, 1972.

"The top!" he yelled down to me. Glad to move after an hour of inactivity, I struggled up the vertical wall, scrambled through the hole, and joyfully joined him on the card-table-size summit. The world was dazzling, ranges of peaks filling half the sky. I was elated to think that we were standing on a place on this planet no one had ever been before.

Meanwhile, the other six climbers caught up with us. One by one I belayed them up the steep slope and through the cornice until all eight of us crowded together on the summit. Despite our differences in background, language, and culture, together we had made the first ascent of Tuliyan.

Our reward was to be the one small candy bar especially allotted for the summit.

"Who has the candy bars?" Dave asked. "I've been looking forward to mine for days."

"I think they're back down in the meadow," Ashraf said.

We couldn't believe it. How could we have left the candy bars behind? And then I realized that chocolate was not a part of the traditional Kashmiri diet. The Kashmiris passed around hard-boiled eggs and we toasted our success with the tea in our water bottles. Dave led the descent back to the meadow, where the four of us enjoyed the lush grass, flowers, and the long-awaited candy bars, while the Kashmiris made chapatis and tea.

Back in Srinagar, our colleagues from Tuliyan kindly shared some very valuable information: the location of three challenging, unclimbed ice peaks. First ascents are to climbers like Olympic gold medals are to athletes; it was very generous of them to tell us about the Sheshnag peaks. These peaks—Brahma at 16,300 feet, Shiva at 16,500 feet, and Vishnu at 16,700 feet—are named after the three major Hindu gods and are located on the pilgrimage route to the holy cave at Amarnath. This cave, the Kashmiris told us, contains a huge ice stalagmite whose size changes with the waxing and waning of the moon. Devout Hindus believe that the ice formation represents the holy lingam of Shiva, the god of destruction. During the full moon of August, as many as a hundred thousand Hindu pilgrims journey for five days to the cave to see this remarkable lingam.

Neither Joel nor I had a clue what a lingam was. Back at our houseboat, I approached Yousef, the boat's owner. "What's a lingam?" I asked.

"You can hike up to the temple on the hill to see the lingam of Lord Shiva," he said, pointing to a steep hill nearby. I thought Shiva's lingam was at the Amarnath cave, but Joel and I hiked up the hill anyhow. On top, we saw a three-foot-high oval rock standing on end with a circle of flowers around its base. Squatting in a trance across from the stone were two wandering sadhus, holy men seeking enlightenment, with long, wild hair, wearing only orange loincloths.

Returning from the temple, I asked Yousef again what a lingam was. He looked at the ground and mumbled, "It is the male sexual organ, memsahib."

On June 4, the four of us and three local porters left Phalgam to attempt the first ascent of the Sheshnag peaks. The trail was deserted except for a lone sadhu hiking through the snow in plastic sandals. Our porters uncomplainingly walked on snow-covered trails in shoes made of braided grass.

At Sheshnag Lake, the porters left us at a shelter that was partially collapsed from the weight of the previous winter's record snowfall. Three symmetric peaks with steep fluted faces looked down at us from above the lake—the Sheshnag peaks.

The next day we hefted our sixty-pound packs and slogged up in the stifling sun. Looking for a flat area on which to pitch our tents, we came upon a snow-filled crevasse at 13,600 feet. We were undecided about camping in the crevasse when Joel threw down his pack and pronounced, "This campsite has a unique feature possessed by no other campsite anywhere in the world."

"What?" I asked. The campsite didn't look that special to me.

"We're here!" Joel said, shaking our tent out of its sack.

At dawn we began picking our way through a mass of crevasses and seracs in the icefall above our camp. Progress was excruciatingly slow because of the worst possible snow conditions: brittle crust over deep powder. I'd put my foot on the hard surface, break through into calf-deep thick powder, pull my foot out of the jagged hole, and start all over again. For hours we trudged painfully toward Shiva, the central Sheshnag peak. Falling through the breakable crust was so unpleasant that Toby and Dave decided to turn back. Joel, who was roped to me, started to follow them down. But I wanted to continue high enough to see the unknown valleys on the other side of the ridge leading to the peak. "Cut the rope, Joel," Toby suggested as he and Dave headed down. We all laughed, although I wasn't sure if this suggestion was completely in jest.

Joel did not cut the rope. As we fought our way up, I didn't think about how exhausted I was or how bad the snow was. Fueled by my curiosity, I was determined to reach the ridge. It was the old step-and-breathe routine: breathe many times, step forward on the crust, break through and sink to my knee in the snow, breathe many times, pull my foot up out of the hole and step again. Sometimes I crawled on my knees, which kept me from sinking down as far. Plodding upward I remembered swimming endless mind-numbing laps on the swim team at the Y.

After breaking trail through the heavy snow for hours, I suddenly found myself on steep, hard ice, the kind that delighted Joel and Toby but frightened me. Joel took over the lead and I followed him up the 50-degree face, with huge flutings of protruding ice. My heart racing and my mouth dry, I continued up a ridge that turned out to be more than 60 degrees.

Rappelling after our first ascent of Shiva. Kashmir, 1972. *(Photo by Joel Bown)*

Seeing that we were heading for the top, Toby and Dave raced back up the broken trail to join us. Then Toby and Joel alternated leads for two hours of intense ice-climbing. Although I felt out of my element, I struggled along behind them. Reaching the top at last, we gave a loud cheer and I flew my tRNA flag from Shiva, our second first ascent. Our magnificent view of the forbidden high plateau of Ladakh, the beautiful twin ice pyramids of Nun Kun, and ranges of unknown peaks was reward enough for our effort. Someday, I thought, I will climb there. Our celebration was cut short by black clouds

(From left) Toby, Dave, me, and Joel relaxing in chairs from a pilgrim rest house after making ascents of Brahma, Shiva, and Vishnu, the three peaks behind us. Kashmir, 1972.

and peals of thunder. We hung our rope around a convenient rock and rappelled down the steep ice face in the lightly falling snow. Back at our camp in the crevasse, my three companions congratulated me on persevering when they'd been ready to give up.

"You're the Queen of Tenacity, Arlene!" said Joel. "I would never have guessed there was a solid slope beyond that terrible snow. I'm glad you went on—and that I didn't cut the rope."

Then Toby spoke. "Arlene, I appreciate all you do to make our trip work: getting up in the dark to cook breakfast, melting snow, and feeding us all. We never would have made this climb without your fighting your way up for hours today. Awesome."

I thanked him, delighted at how supportive he sounded, and optimistic that this could mark the beginning of a better relationship between us.

The next day, Joel and Toby climbed ice to the top of Vishnu while Dave and I slogged through knee-deep snow breaking trail to the easier summit of Brahma. We had made the first ascents of all three mountains!*

On a roll, we were ready for the next part of our trip: Bhutan. Just before we'd left home, I'd learned that the isolated kingdom of Bhutan had joined the United Nations and was going to admit tourists for the first time. I wrote the Bhutanese, asking for permission to climb. I was overjoyed when they replied that they had hundreds of unclimbed peaks and would welcome our visit. We were the first to inquire and the first to get permits to climb in this mysterious kingdom.

We returned to Srinagar from the Sheshnag peaks to discover that India, which controlled access to Bhutan, had revoked our permit because the United States had sided with Pakistan in the recent war over Bangladesh.

Disappointed to miss Bhutan, we now had an extra month for a new adventure. Our Kashmiri friends, impressed with our four first ascents, told us about the Kishtwar Himal, a range of peaks that were so remote, steep, and difficult that few climbs had even been attempted. Foreigners were officially not allowed into the area, but it was so far off the beaten track there were no checkpoints. The Kashmiris gave us a rough sketch map and a letter with an elaborate purple wax seal that we could present in case anyone happened to question our presence.

A few days later we loaded ourselves and mountains of climbing gear into the overcrowded daily bus to Kishtwar. The bus had been built for small passengers, and its ceiling was so low and the seat ahead so close we felt like giant pretzels as we contorted into our seats. Thirteen hours of bone-rattling travel later, we stumbled out of the bus and found ourselves in a tidy mountain town with spring in

*Waiting out a storm in a New Zealand hut some months later, I happened to read *Twenty Years in the Himalaya* by Colonel G. Bruce. To my surprise, the book included a photo of Brahma, Shiva, and Vishnu and a description by Colonel Bruce of returning from the Wardwan Valley to Kashmir in 1898 by traversing the northern peak. It is likely that Bruce ascended Brahma, but not Shiva or Vishnu, where the easiest routes require climbing ice that is steeper than 50 degrees.

Joel surveys the precipitous unclimbed peaks of the Kishtwar Himal. Jammu, India, 1972.

the air. Brilliant wildflowers were blooming in the meadows and pairs of long-tailed magpies whirled by in ecstatic mating flights.

Settling into a comfortable government rest house, we asked about the trail to the Kishtwar peaks, but no one in town had heard of any mountains in the area. We continued asking everyone we met for two days. Just when we began to doubt the existence of these mountains, Mr. Mallik from the Jammu and Kashmir Mountaineering Club serendipitously arrived to begin a new job in Kishtwar. He found us two horsemen and horses to help carry our gear and then showed us the trail out of town.

For the next four days we followed a rugged, steep track above the Maraw River that was supposed to lead to the Kishtwar Himal.

Nomad caravans, elegantly veiled women with fine gold jewelry, mule trains heavily laden with bags of grain, and men with briefcases and umbrellas, looking as though they were on their way to the office, all amicably shared the path. To our surprise, the trail was bordered by large expanses of a tall, leafy green grass of the sort that would have been most popular in Berkeley.

In my faltering Hindi, I asked locals about the way to the Kishtwar Himal and was met with silence or a shrug. Finally, on the fourth afternoon, we reached the end of the trail, a green meadow at about 11,000 feet jutting up against a rocky moraine with clouds boiling all around. We went to sleep, wondering if we had spent the last week on "a wild mountain chase."

We awoke in the morning to an overwhelming view of the Kishtwar Himal towering two vertical miles above us. Spectacular peaks filled our eyes and our minds to the point of utter bewilderment. A steep pyramid of rock and ice rose unapproachably to over 20,000 feet.* A sheer granite face thrust 6,000 vertical feet above the glacier, somewhat like El Capitan in Yosemite Valley, but twice as high. All the routes required multithousand-foot ascents on high-angled granite and ice. The granite looked smooth, the ice hard, the glaciers broken up, and the cornices menacing. We'd found a feast of mountains, but we didn't have the appetite or the utensils for the meal.

"Let's just stay down here and take pictures," I gulped. "No way we can get up those."

"Small wonder these peaks are unclimbed," Joel agreed. "Well, at least we came and we saw. Now it's time to flee."

Just then the clouds boiled up and overflowed, and rain poured down on us for the next three days. On the fourth morning Joel's jumbe bracelet from Africa turned black—an inauspicious omen.

"Kishtwar was not on the Endless Winter itinerary," Dave said. "You can see that straying from the plan is trouble, trouble, trouble."

---

*We later learned this pyramid in the Kishtwar Himal is named Sickle Moon. At 21,571 feet, it is the highest peak in the range and was first climbed in 1975 by an Indian Army expedition from the High Altitude Warfare School.

Everyone agreed it was time to go. I suggested an alternative route back to Kashmir that would avoid the dreaded thirteen-hour bus ride. From our sketch map it looked like we could trek straight back through the Wardwan Valley in four days. We packed up to walk back to Srinagar, not considering that this route wasn't on our itinerary either.

The first night we stayed in a cozy guesthouse built for British officers in the nineteenth century. Signing the guest book, we saw that relatively recent guests included the famed French climber Claude Kogan and her party on their way to climb Nun Kun—nineteen years earlier.*

The continuing downpour turned the trails into bottomless mud. On the second morning, our horsemen announced they were turning back, and dropped our gear on the ground. "We're slightly overequipped for trekking," Joel noted as he distributed 150 pounds of ropes, pitons, ice screws, and other technical climbing gear for the four of us to add to our packs, now an oppressive eighty to one hundred pounds each.

The remainder of our retreat to Kashmir was memorable for mud, misery, and bad food. The straight-line distance on the map had looked reasonable, but the map hadn't shown that the trail went up and down thousands of feet to every house along the valley. We were starving for a decent meal, but there was no fresh food available. All we had to eat were unpalatable tins of decomposed Indian sardines that appeared to have been in the bazaars since the days of the Raj. Finally we bought an ancient chicken from a villager and boiled it for hours, hoping it would soften up enough so we could eat it, but it only

*Claude Kogan climbed Salcantay in Peru with my Reed advisor, Fred Ayres in 1952. In 1953, along with Swiss missionary Pierre Vittoz, she made the first ascent of Nun, the highest summit of the beautiful twin peaks we'd seen from the summit of Shiva. The next year, Claude set her own women's altitude record, reaching 25,600 feet on Cho Oyu in Nepal. She returned to Cho Oyu in 1959, leading the first-ever women's expedition to an 8,000-meter peak. Claude Kogan and Claudine van der Stratten were poised to reach the top when a series of intense blizzards followed by a lethal avalanche struck their High Camp. Both women and two of their Sherpa porters perished, ending for many years the idea of a women's expedition to 8,000 meters.

turned into chicken-flavored gum. Endlessly, we fantasized in mouth-watering detail about the dinner we would prepare when we returned to the United States. It grew to twenty gourmet courses, ending with a cake in the shape of the Parthenon covered with an inch-thick layer of frosting.

Our last obstacle was a 15,000-foot pass back to the Vale of Kashmir. When we finally made it over the pass with our huge loads and reached the first rest house, I collapsed on the porch and slept until morning wearing all my clothes, too tired to eat or get into bed.

The next day we passed a farmer selling bags of luscious red Kashmiri cherries. They looked so good that Joel and I gobbled down a pound each without washing them. Cherries grow on trees, I thought, they're fine.

Then we saw another vendor take a handful of cherries, bend over, wash them in the gutter at his feet, and put them in a bag for the next customer. This gutter water was all-purpose: drinking, dish-washing, and sewer. Just looking at it made Joel and me feel sick.

Back in Srinagar, Yousef, our houseboat owner, invited us to a traditional meal at his family home. Seven men and I sat on the floor and enjoyed a dinner of tasty Kashmiri curries served by three graceful, veiled women.

I asked Yousef if the women might join us. "They will eat after us, what is left," was his firm reply. Seeing my shocked face, he explained that this was their way, and they were all used to it. "The women enjoy seeing you eat the food they prepared. Eating with you would not be comfortable for them."

I came to understand that we were guests of people whose customs were different from ours and it was up to us to adapt. In the past months, I'd learned to wear more modest clothing, be patient, and listen carefully to avoid misunderstandings. I had realized the importance of loyalty to guests, friends, and family. Our being accepted as honored guests of the Kashmiris led to extraordinary hospitality, including their telling us about their unclimbed peaks. If women did not eat with men here, it was not my place to protest.

After dinner Yousef took me to the area for the women and children. Yousef's mother, the respected matriarch, sat on a string bed giving orders while the younger women offered me sweet, milky tea. The now-unveiled women were sharing the food that had been brought back from our dinner, giving the choice morsels to the old woman and the children.

Yousef's wife, Fatima, stood to greet me with a shy smile. She was young and attractive with large gold hoop earrings, a gold nose ring, and a brightly colored silk tunic over her loose trousers. Yousef explained that the women were curious about me and happy I'd come to talk with them. "They see you as a different sort of creature, neither man nor woman. You eat dinner, talk, and climb mountains with the men, but you look like a woman."

Zeena, a sparkly twelve-year-old girl, was eager to practice her English and interpret for me. I asked her to thank the women for the delicious dinner. They giggled and whispered among themselves.

"My aunties want to know where your children are," she translated.

I thought quickly. "Joel and I are only just married and have no children yet."

"We all wish you many strong sons."

There was a flurry of conversation in Kashmiri, and Zeena shyly asked me if I could look at her grandmother's foot. I went over and the old woman pulled up her baggy silk trousers to reveal a foot twisted with what appeared to be crippling arthritis. I shook my head sadly. I couldn't help.

One of the women finished pounding spices and poured them into a teakettle. I accepted a second cup of tea, now with a sweet, spicy flavor. As I played peekaboo with the young children and bounced the chubby baby on my knee, I felt a bond with these Kashmiri women whose lives were so different from mine. And I realized how fortunate I was to be able to inhabit two worlds—that of men, and also that of women.

Chapter 14 ▲

Sharing a Room with My Mom

Chicago 1957

I come home from school and make my way through the smoke and the pungent smell of chicken fat to the small bedroom I share with my mom. I close the door.

The walls are painted sky blue, my favorite color, and the curtains and dust ruffles are a lighter shade of the same blue. The small closet in the corner overflows with our clothes and my mother's sheet music, violin, and records—remnants of her career as a violinist. Her double bed with heavy oak bedposts fills a windowless corner. My single bed is beneath a large window; my pillow is often covered with fine soot particles that float over from the steel mills in nearby Gary, Indiana.

I'm lying on my bed reading The Count of Monte Cristo *when Mom comes home. And talks to me.*

"After I pay our bills and give this month's rent to Grandpa, I have seven dollars left. Do you want to go shopping? I need a new dress for work, but have no idea what to buy."

"You should be able to buy your own dresses by now, Mom," I say, not bothering to look up from my book. "I've been picking your clothes out for years. And you can't get a decent dress for seven dollars."

"But you have such good taste." My mother's tone is conciliatory.

"I do need a birthday present for Carol's birthday," I say. "She's turning twelve, like me."

"How about a portrait of Bach? He's the greatest composer—"

"Are you kidding?" Out of the corner of my eye, I see my mom fishing pieces of paper out of the wastebasket, but I don't say anything.

"Well, you can have three dollars to buy Carol anything you like. Just don't tell Grandma or Grandpa."

"It's none of their business."

Mom smoothes the creases out of a crumpled piece of paper. "What's this? You wrote to someone named Sue that you want to run away from home. Who is Sue? Is she your friend from camp? What's going on?"

"Stop snooping. You shouldn't read my private notes."

"It was just sitting there. Grandma and Grandpa want only the best for you. You should do what—"

"They're always yelling at me and you're always prying. You're all awful. I can't take it anymore." I close my book and sit up.

"Grandma and Grandpa are loving and protective. They know best."

"They don't know anything. Why do we live with them anyway?"

"They cook the best food for us and they take such good care of us. And you know I don't earn enough to support us."

"Someday I'll get a job and we can get away from them."

"Not now," she says. "You should read, study, get a career of your own so you can always take care of yourself. I married a doctor, just like Grandma told me to, and look what it got me. And now I'm supposed to get myself another husband."

Contrary to my grandparents' hope that my mother will find me a new father, she rarely goes out with men. When she does, she returns with a detailed litany of her date's faults.

"Men are like babies," she says. "They go on and on talking about themselves. They're only interested in one thing. If you go out with men, be careful."

Her words make little sense to me, so I ignore her as usual.

"Remember the TV show we saw where the man murdered his wife? That's what men do."

Although I pay no attention to my mother's warnings about men, I'm happy to heed her advice to read. Through books, I escape to another time, another continent, another planet, another family. I make a plan to read all the books in the library in alphabetical order.

On Saturdays my mother and I go to the library and then to the movies. After buying our tickets and seating me, she asks if she can be excused.

"Why?" I ask.

"I need to relax," she answers. "I'll be back soon."

She always returns before the movie ends. She tells me she's been window-shopping and relaxing with a cigarette and coffee in a café.

When I ask her if I can invite my friend Carol to come to the movies sometime, she says no.

"I'd rather we have our Saturday afternoons together, just you and me."

The Endless Winter in Afghanistan and Nepal ▲

1972

IN JULY 1972, Joel, Toby, Dave, and I flew from Kashmir back to Kabul, Afghanistan, on Ariana Afghan Airlines, a carrier that Joel referred to as "Scariana Airlines—three frights a week." Our objective, Noshaq, at 24,450 feet, was the second-highest mountain in Afghanistan and by far the tallest peak we'd ever attempted. It was located at the gateway to the Wakhan Corridor, a little-known and strategic strip of Afghan land bordered by the Soviet Union, China, and Pakistan. Few Westerners had ever been allowed to visit this politically sensitive area, and we felt lucky to have permission to attempt Noshaq.

Arriving in Kabul, we learned we weren't the only lucky ones; the Afghans had given Noshaq permits to a half dozen other expeditions. At the same time, our climbing family grew to ten. We were joined by Sue Darling, a British climber we'd met in Kenya; Joel's climbing and skiing partners Dave and Annie George from Salt Lake City; and my Denali teammate Margaret Young, along with two of her friends, Bill Griffith and Earl Fernham.

In Kabul, a facade of modern buildings hid a maze of dirt streets and mud-walled homes. Crowded bazaars alternated with cultivated fields, poplar groves, and parks. In the carpet area of the bazaar, precious rugs were laid across the roads, bearing the traffic of people, animals, and cars; a carpet merchant explained that wear mutes the colors in a rug, increasing its value.

Inside the bazaar, men carried piles of snowshoe-size flatbreads on their heads; others were bent double by huge burlap sacks of grain on their backs. Afghan women, hidden behind tentlike *chadris* or wearing high heels and Western-style dresses, bargained at stalls displaying beautifully arranged fruits and vegetables.

Our first task was to purchase a month's worth of food. Once again edible packaged food was elusive. We could tell the antiquity of the American food by the lack of zip codes—instituted nine years earlier—in the addresses on most of the labels. Opening a can of Camembert to check the quality, I stepped back as the cheese erupted three feet straight up like Vesuvius. We eventually managed to buy what we needed, only to discover later our purchases included substances we nicknamed "Weevilos," a breakfast cereal complete with crawling insects, and "Gasolinos," Cheerios that had been stored next to barrels of gasoline.

Merchants often invited us to sit and chat in the backs of their stalls. Sipping cardamom-flavored green tea with a carpet seller, I learned that Russia and the United States were competing for influence in Afghanistan. The merchant smiled broadly as he told us about a grain-storage silo in Kabul built by the Russians and filled with wheat from the United States. One market was named the Nixon Bazaar because of the American used clothing sold there.

We learned from a jewelry merchant that the king, Zahir Shah, supported equal rights and education for women, and had gone so far as to tell his ministers their wives must unveil to set an example. This merchant told us he supported the king's reforms, but many traditional Afghans still followed a patriarchal, tribal way of life and were threatened by change. But people were tolerant of foreigners and no one bothered me whether I wore jeans or a skirt.

Our supplies and gear finally organized, the ten of us plus an interpreter and several new friends from Kabul headed for the Wakhan Corridor. We bounced over a rough, potholed dirt road just south of the Oxus River, which separated Afghanistan from the Soviet Union. Just across the river, Soviet military caravans sped along a paved road past tidy Soviet towns. It was the height of the Cold War, and after years of indoctrination about the evils of the Soviet Union, the "enemy" seemed ominously close.*

Nonetheless, we peacefully motored past fields of bright purple and white opium poppies that stretched as far as we could see. This valuable and illicit crop was a likely reason foreigners usually weren't allowed into this area.

We had to stop frequently for bathroom breaks, as Joel and I suffered from acute amoebic dysentery, a probable result of our cherry feast during the rout from Kishtwar. Returning from such a stop, I noticed Toby negotiating for some of the local crop and wondered if his interest in such substances contributed to his distant demeanor. In any case, having so many people in our group diluted the tension between us.

Stopping for lunch in a green meadow, we were joined by a tribe of Kuchi nomads traveling with their herds of camels, sheep, and goats. The women wore dresses with heavily embroidered bodices decorated with silver coins, and flowing skirts made from yards of bright red, orange, and purple cotton. The nomads were amazed to

---

*Shortly after our visit to Afghanistan, a coup against King Zahir Shah brought a series of left-wing governments to power. Although they promoted universal education, land reform, and rights for women, they were dictatorial and many Afghans, especially the more traditional, resisted their rule.

Unable to control the rebellion, the beleaguered Afghan government requested help from the Soviet Union. In 1979, Soviet tanks crossed the Oxus River and rolled through the streets of Kabul. After overthrowing the existing government, the Soviets installed another Communist regime. For the first time in its history, Afghanistan was subject to foreign domination.

The United States, in its Cold War mentality, supported the Afghan resistance, including the most reactionary of the mujahedin, in spite of their extreme ideology. Thousands of volunteers, including Osama Bin Laden, came from throughout the Muslim world to help the Afghans defeat the Soviets. The United States provided them with millions of dollars' worth of weapons and training, channeled through Pakistan's military intelligence agencies.

Thus began decades of civil war and suffering for the Afghan people. Defeated by the mujahedin and on the verge of splitting apart, the Soviet Union withdrew from the country in 1989. The ensuing competition among mujahedin parties led to chaos. The anarchy was finally contained by the Taliban, who imposed a tyrannical fundamentalist regime on the Afghan people. The Taliban were removed from power by the United States in 2001 following the terrorist attacks of September 11, 2001.

After sharing our lunches, the Afghan children were politely curious about us.

meet Westerners—especially women wearing jeans. More open than the veiled urban women, the Kuchis asked us about our families through our interpreter and shared their simple lunch of goat cheese and hot bread.

Arriving at the dozen mud-walled huts that comprised the village of Quasi Deh, we hired porters and enjoyed a pleasant two-day trek through the arid countryside to Noshaq Base Camp. Though we'd expected some company, we were surprised to find a small city of tents pitched on the moraine below an enormous glacier. The famed Himalayan climber Reinhold Messner was there, guiding a group of forty-nine Italian climbers and their twenty-two Sherpas. Messner and his wife inhabited a palatial tent complete with curtained windows and stone walkways. Germans, Norwegians, and Poles also populated Base Camp, which Dave Graber christened Tentiopolis.

While we set up our camp, Wanda Rutkiewicz, a Polish climber whose shiny black hair framed broad cheekbones and a dazzling smile, stopped by to introduce herself. Sharing stories with the warm and gracious Wanda, I learned she had climbed the hardest north faces in the Alps and hoped to lead a steep technical new route on the far side of Noshaq. She told us that her team, too, included four women.

As expedition after expedition toiled past with heavy backpacks, our team relaxed in the sun, melted snow for water, cooked and ate large meals, and contemplated the formidable peaks surrounding us. We were testing an innovative new technique for high-altitude mountaineering: sloth acclimatization.

Part of the reason we were taking it easy was that Joel and I were still recovering from our dysentery; the medicine that killed the amoebas made us lethargic and diverted our attention to our guts. Like many Westerners living or traveling in Asia, we had a morbid fascination with our intestinal processes. As we acclimatized, we entertained ourselves by developing a graphic nomenclature for our excretions, ranging from a dreaded "aerosol" to "butterscotch pudding" to "Sheshnag applesauce" to a "chocolate chip with a handle," culminating in the highly esteemed "city dump."

Due perhaps to his dysentery, Joel became infamous for the amazing odors he emitted. Proper Sue Darling never had this problem. However, one afternoon, an especially foul smell emanated from Sue. We all stared at her in surprise and started moving away from her side of the tent. Sue looked bewildered.

"It's still Joel," Dave quipped. "He's an anal ventriloquist."

After several days of sloth acclimatization, we were ready to climb our first peak, 18,688-foot Korpusti-Yaki. We sank into deep snow softened by the hot sun—to our calves, our knees, even our thighs—but struggled on to the summit. Once there, my tRNA flag in hand, I was rewarded with the enticing sight of two forbidden mountain ranges, the heavily glaciated Pakistani peaks in the south and the arid Pamir range in the USSR to the north.

On the way down we surprised a young Italian couple in skimpy beach attire kissing on a rock in the middle of the glacier. In a country where the genders rarely mingle and extramarital sex can be punished by death, this was an extraordinary sight.

A straightforward, three-day ascent of Aspi-Safed (21,300 feet) provided us with striking views of our objective, Noshaq. Back down at Base Camp, Reinhold Messner told us that he and his three Italian clients had made it to the summit. While we were away, Margaret Young, Bill, and Earl had also attempted Noshaq and then headed back to England.

We asked Messner if he knew whether Margaret and her companions had reached the top. "You Americans do not have the alpine experience needed to climb such mountains," he told us, going on to explain that Noshaq was too high and cold for a woman of Margaret's age. Indeed, two of his clients and he, himself, had gotten frostbitten on their ascent.

Later that day, a Polish climber brought us a note from Margaret. She wrote that she and Bill had easily reached the summit.

I was surprised that forty-year-old Margaret had climbed Noshaq in two weeks without any problems, while Messner, one of the world's strongest climbers, had suffered frostbite. My guess was that the Europeans were not drinking enough liquid, essential to prevent frostbite and altitude sickness. The Italians depended on their Sherpas to melt snow; their liquid intake was a few cups of tea each day. We Americans drank many quarts of water each day—a powerful technique for safe high-altitude climbing that few climbers appreciated at the time. The experiences of Messner and Margaret on Noshaq strengthened my belief that for an expeditionary climber, the ability to melt ice can be as important as the ability to climb it.

Meanwhile, Annie George developed tendinitis, so she and her husband left to visit Bamiyan, Mazar-e Sharif, and other cities in Afghanistan. Then Toby abruptly announced that he was joining the Poles to attempt a more difficult route up Noshaq. In spite of our recently improved relationship, I was relieved.

Finally Joel, Dave, Sue, and I were ready to try for the top of Noshaq. After choking down an oversweet breakfast of cocoa and Pop Tarts, we hefted sixty-pound packs to begin a week of carrying loads to equip three camps on our way to the summit.

The day we moved our camp to 23,000 feet, higher than I'd ever been before, was a tough slog. We spread out, each going at his or her own pace. As I moved steadily up a thousand feet of loose rock in lightly falling snow, a shadowy figure emerged from the mist at the bottom of a rocky cliff above me. It was Wanda Rutkiewicz, triumphant on her way down from the summit.

Wanda gave me a warm hug. "We women climbed to 7,500 meters," she said. "Now we go to 8,000 meters, all women."

Her words stayed with me as I focused on getting up a steep cliff band using a fixed line left by a previous party. Carefully, I moved

Wanda Rutkiewicz and companion at 22,000 feet on their way down from the summit of Noshaq in 1972. Wanda suggests we colead a women's climb to 8,000 meters on Annapurna I.

across a short exposed rock traverse. A misstep could mean disaster. When I reached a gentle snow slope, I relaxed and my thoughts returned to Wanda's words.

Only fourteen mountains in the world soar above 8,000 meters (26,250 feet). The first successful ascent of a peak of that height was in 1950, when Maurice Herzog's French team scaled Annapurna I, the tenth-highest mountain in the world. Since then, all the 8,000-meter peaks had been climbed by men; not a single woman had reached that altitude. Eight Polish, American, and English women were now attempting to climb to 7,500 meters on Noshaq, and most of us had already reached 7,000 meters. After Denali and Noshaq, I had little doubt that women could climb to 8,000 meters. Maybe if Wanda and I joined forces, we could give a team of women the opportunity to try one of the world's highest mountains. The vision gave me energy as I trudged upward through deep snow, loose rock, and high wind.

We arrived at High Camp cold and tired. The tents, sleeping bags, and other gear we'd been carrying around the world were by necessity lightweight and most appropriate for climbing in the tropics. It certainly was not tropical here. All night the wind blew, the snow fell, and we froze.

Low temperatures and high winds greeted us on summit morning as we reluctantly emerged from our tent. Hunched over against the gale, we fought our way upward, one freezing step at a time. When we reached 23,700 feet, about 800 feet below the top, Dave's fingers were hard and white, so he decided to turn back.

My toes and fingers, sensitive ever since Denali, ached in the bitter cold. But we had come so far and the top was tantalizingly close. Each step took me higher than I'd ever been and I longed to explore the unknown physical and psychological terrain that lay above. But I knew that in this high, hostile place, living creatures had a tenuous existence and the line between life and death was narrow.

I had to make the wrenching decision whether I should continue upward with my fingers feeling like ice cubes, or listen to the prudent voice telling me to go down. To climb high on a big mountain requires months or even years of effort: selecting an objective and a team, fund-raising, travel, carrying heavy loads. And after this enormous effort, just before the top there always seem to be good reasons to turn back. It's usually too cold, too steep, too dangerous. To succeed in an ascent, I often had to persevere through fear, fatigue, and technical problems to complete the final slopes and reach the top. On the other hand, most accidents happen on the descent, so the wise climber must decide when to push on and when to turn back—and live to climb again.

Although I thought I could make the top, I wasn't as sure I could get down safely. The summit wasn't worth risking frostbite or worse. Thinking about my friends, my cats, and my life back home, I yelled to Joel and Sue, resolutely heading upward, and pantomimed that I was going down. Then I cast a last longing glance at the summit, turned around, and ran back down, following my footsteps in the snow until I caught up with Dave. A little later Sue also decided to turn back and she soon joined us. Together the three of us headed back to the relative warmth of our tent.

Meanwhile, Joel struggled on alone to reach the 24,300-foot middle summit at midday. He told us that it had taken him two hours to climb the last 600 feet and that the gale was so fierce on the summit that he couldn't stand upright. All he could see was a cloud of fog.

Later that afternoon, after the wind dropped, four Polish climbers also at the High Camp left for the top. The next day, they told us they had reached the main summit near dusk, but didn't get back until 2:00 A.M. One of their team, Alison Chadwick-Onyszkiewicz, a British artist married to a Pole, had lost her crampon in the dark below the summit, slipped, and fallen several hundred feet. As the others watched in horror, she managed to stop herself just above a fatal drop-off.

We were surprised that the Polish climbers had headed for the top knowing they would have to descend in the dark. But they were part of the world of semiprofessional alpinists whose jobs and lives were climbing. Delighted to leave their Soviet-controlled country for this expedition, they were strongly motivated to succeed so they could go outside Poland again. We, on the other hand, were recreational climbers with the luxury of turning back if the conditions were too miserable or the danger too great.

After everyone returned safely, we held a grand celebration at Base Camp. Our team contributed a half dozen unique desserts made from our remaining edible food: chocolate bars, dried fruit, and Grape-Nuts. The Poles made a huge meat-and-potato salad and shared the treat of all treats: tinned strawberries with sweetened condensed milk. They then passed around Russian vodka and caviar and sang Polish freedom songs.

Wanda called Sue and me over to pose for a photo with her and the other three women on the Polish team. Standing with these strong, beautiful women, I felt a warm, loving sense of having many sisters.

"Women can go even higher than this," Alison said. She thought there were several Polish women who could climb to 8,000 meters. She turned to me and asked, "Do you know of others?"

"Six of us climbed Denali and I expect there're lots more who'd love to come," I said. I imagined an international team of women sharing the hard work and the risks of the expedition. It was an audacious idea that a group of women who didn't even speak the same language could climb that high together.

"Yes, we climb." Wanda articulated carefully. "To the top. Together." I felt a rush of adventure and also of fear. I knew that 8,000-meter peaks were much more dangerous than the mountains I'd climbed in the past. It was hard not to dwell on the chilling statistic: for every ten climbers who attempt such a peak, one does not return.

Wanda, Alison, and I exchanged addresses, agreeing to stay in touch.

Back in Kabul, we were starving for good food. Dave and Joel managed to score a gallon of ice cream and a large jar of Hershey's chocolate sauce from the American Embassy Commissary. They cut the ice-cream block in two, poured half the jar of syrup over each portion, and devoured gargantuan chocolate sundaes. The rest of us gorged on succulent roast lamb kabobs, naan (a delicious hot bread baked in a tandoori oven), and *kharbooza,* the crisp, white-fleshed Afghan melons that are said to be the best in the world.

Well fed and energized, Joel and I said good-bye to Toby and Dave Graber, who were planning a separate Nepal trek. I would miss good-natured Dave, but once again felt grateful to be free of Toby. We had reached some measure of peace before going our separate ways, but I still didn't understand his negativity toward me.

The next day, Joel, Dave and Annie George, and I, along with Tom Stephens and Bill Conrod, two more of our good friends from Salt Lake City, flew to the Nepal Himalaya, the world's most magnificent mountains. We planned a thirty-day trek that would take us over the 19,000-foot Tesi Lapcha Pass, up some peaks near the pass, and then to the base of Mt. Everest.

Intended to be the climax of our trip, our time in Nepal was disappointing. The weather was stormy, the Mt. Everest area overflowed with trash and tourists, our head Sherpa made off with much of our food and supplies, and the cook's hygiene was questionable—with the inevitable intestinal results.

The low point was a blizzard-filled week spent camped in a slushy potato field above the village of Khumjung at 12,000 feet, all six of us racked with dysentery and colds. The snow was so heavy our tent pole broke, rendering everything sodden and freezing. Appropriately, I was reading *The Gulag Archipelago,* a Solzhenitsyn novel set in a Siberian prison camp. Although we were in our gulag by choice, I could readily relate to the wretched physical conditions endured by the prisoners. Although each day seemed wetter, colder, and more miserable than the last, we were cheered by the knowledge that Mt. Everest was very close. When the storm finally ended, we hiked up Kala Pattar. From the 18,200-foot summit, we admired the awe-inspiring view of Lhotse, Nuptse, and Mt. Everest. The proud black pyramid of Everest, with its plume of jet stream–blasted clouds, looked even more unattainable than the white towers of the Cascades had years earlier. The thought of trying to climb the world's highest mountain drifted into my mind, but I let it go as implausible and it vanished into the dark blue high-altitude sky.

Our Salt Lake friends had to go home, and Joel and I had new environments to explore. After a few days of happy gorging at the food stalls on the island of Singapore, we headed for the underwater wilderness of Australia's Great Barrier Reef. Arriving in Cairns in northern Queensland eager to go snorkeling, we were dismayed to discover that the reef was many miles offshore and the only way to get there was by boat. After a choppy two-hour ride, we jumped into the warm water and dropped down into the colorful world of fish and coral. Instead of working against gravity to climb mountains, we floated, weightless, and ascended vertical walls by merely kicking our legs.

After two weeks in Australia, to our mutual regret, Joel had to return to the United States. We hadn't fallen in love, but we knew we'd be lifelong friends. Together we'd followed most of the route I'd envisioned spinning the world globe in the chemistry library after seeing *The Endless Summer.* At a time when U.S. travelers rarely ventured outside Europe and the Americas, we had found our way around the world carrying out expedition after expedition in some of the most spectacular mountain ranges on earth. We had made three first ascents and had climbed higher than 23,000 feet. (It was only later that I realized we had also had the very good fortune to climb in the mountains of Ethiopia, Uganda, Iran, Kashmir, and Afghanistan shortly before they became inaccessible to foreign climbers.)

After hugging Joel good-bye, I flew to New Zealand, where I met up with my old Denali friends Margaret Clark and Faye Kerr. After a few weeks of climbing with them in New Zealand and then scuba diving in Fiji, Tonga, and Samoa, my Endless Winter came to its end. It was time to go home, which can be the hardest part of a journey.

Dave Graber wrote about our trip years later:

> Imagine taking four people who had not climbed together or been linked by friendship and put them up against: maddening bureaucracies; daily clashes with unfamiliar environments, cultures, languages; difficult climbing and trekking trips in totally unknown, unexplored, unsafe regions; illness, close quarter confinement, poor food or no food; discomfort of every kind, and make them experience nearly every possible emotion isolated as a tiny group in a prolonged timeline living each day facing the unfamiliar. I look back to the Endless Winter as probably the most maturing event of my life.

I shared Dave's sense of growth during our trip. I had found that the intensity of high-altitude climbing, far above the world and its cares, gave me some peace. The St. Elias tragedy, my grandfather's death, and my mother's pain would be with me for the rest of my life, but the passing of time had begun to heal the raw edges of my sorrow.

Nonetheless, my return to California was marked by culture shock and disorientation. Before leaving on the Endless Winter, I'd accepted a postdoctoral position in Seattle so I could climb in the Pacific Northwest. But now my priorities were different; I wanted to stay close to my friends in Berkeley. When I was offered a postdoc at Stanford Medical School to study protein folding, I didn't know where I wanted to be. Racked with indecision, I moved back and forth between Seattle and Stanford with all my belongings—including my two beloved cats—three times! And I still had no idea what I should do.

In Palo Alto, I enviously watched my black cat sleep in a patch of sunlight. The indolent feline knew exactly where she belonged and what she wanted; I so wished I were that cat, and that my only responsibility were to lie in the sun. Finally Buzz Baldwin, my Stanford advisor, gave me some excellent advice: "Don't worry about what you ought to do or what other people will think," he said. "You cannot predict the eventual outcome of your choice. All you can do is listen to that little voice deep inside you that knows what you want to do." And so I picked Stanford.

The environment in the Stanford Biochemistry Department was electric when I arrived in the fall of 1973. Students in Paul Berg's lab had just succeeded in taking a piece of DNA from one organism and putting it into another. These first recombinant DNA experiments eventually led to the field of biotechnology.

My own scientific work at Stanford was trying to understand how proteins—the molecules that perform all the basic functions in our bodies—fold rapidly into exactly the correct three-dimensional structure. Proteins are first synthesized in our cells as long strings of building blocks called amino acids. If a protein were to try all the different folding possibilities before it found its correct structure, it would take longer than the lifetime of the universe. Amazingly, most proteins fold up in less than a second, somehow knowing the correct configuration for their function. Scientists wanted to know how and why this is possible, and I thought finding information about the structure of a partly folded molecule would provide some clues.*

---

*Many serious illnesses that affect the brain, such as Alzheimer's, Parkinson's, Huntington's, and mad cow diseases, as well as other conditions, including some types of diabetes, cataracts, and cystic fibrosis, are caused by incorrectly folded proteins. Information about how proteins fold is vital to research on cures for these diseases.

I was using nuclear magnetic resonance (NMR) to study the protein ribonuclease. The common wisdom of the time was that it was impossible to look at the structure of ribonuclease while it actually was in the process of folding, but I believed I could find a way to obtain an NMR spectrum of the partly folded molecule.

In addition to thinking about how ribonuclease folds, I was running in the hills above Stanford, climbing in the Sierra Nevada, and corresponding with Alison Chadwick-Onyszkiewicz and Wanda Rutkiewicz about a women's expedition to 8,000 meters. My mother was living in her own apartment in Chicago. I was finally settling down to a routine when my life in the United States was once again interrupted, this time by an adventure in the heart of Central Asia.

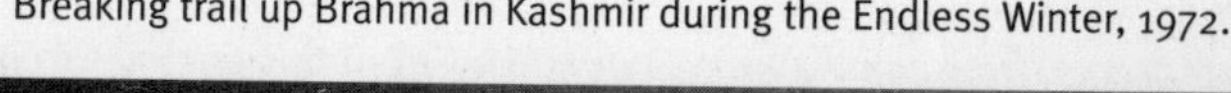

Breaking trail up Brahma in Kashmir during the Endless Winter, 1972.

CHAPTER 15 ▲

Sputnik
Chicago 1957–58

"The Russians have launched a satellite into space," my eighth-grade teacher announces, pointing out the window as we straggle back into our classroom after lunch on a beautiful fall day. Her voice is shrill with alarm as she tells us the satellite's name: Sputnik. Sputnik, *I whisper to myself.* Sputnik. *A foreign and menacing word. Although I'm not exactly sure how it threatens us, I know it's bad that the United States, the leader of the free world, has been bested by the evil Soviet Union. Beyond the red, orange, and yellow trees, I scan the sky for something different and listen for the sound of rockets. How can we protect ourselves from this enemy satellite, this Sputnik of the wicked Russians, flying above us as Miss Austin speaks? During tornado drills, we crouch under our desks and wrap our arms around our heads. During atom-bomb drills, we go out into the hall, duck, and cover ourselves with our notebooks to shield ourselves from the blast and the radiation. I'm not sure what we should do about Sputnik.*

A few weeks later, Miss Austin answers the question. "We will send up our own satellites, bigger and better than Sputnik," she says. "And you students can do your part by learning more science and math."

But I know girls aren't very good at those subjects and am sorry I won't be able to help America win the space race. I have already decided to teach kindergarten when I grow up so I won't have to study much math or science.

Eighth grade comes to an end and I decide to take general science in summer school to get my science requirement over with before high school. But when I go to register, the class is full. I sign up for typing instead.

On my first day of high school, I walk into my homeroom and find a seat in the back. A gray-haired, heavyset woman introduces herself as Miss Mitchell, our homeroom teacher. She tells us that we are the lucky twenty-five freshmen who have been chosen for the Hundred Program.

Miss Mitchell proudly explains that as members of the "Happy Hundred" we have the opportunity to study biology and algebra as freshmen, so we can become scientists and help America beat the Russians.

I must be in the wrong room. I approach Miss Mitchell's large wooden desk. "You've made a mistake about me," I say, my voice trembling. "I'm not a Happy Hundred. I can't do algebra or biology. I'll fail if you make me take those classes. Can't I just take the regular classes like everyone else? Please?"

Miss Mitchell asks my name and checks her list. She looks back up at me with a reassuring smile. "Don't worry, dear," she says. "You did very well on all your aptitude tests. If you work hard, I know you'll do just fine."

Peak Lenin Bares Its Fangs ▲

1974

OFFICIALLY INVITING YOU *International Alpine Camp Pamir from 13 July to 16 August 1974—Sportkomitet USSR.* I read the telegram from the Soviet Sports Foundation and then read it again. I couldn't believe it. I was being asked to take part in a Soviet climbing camp in the high Pamirs, the remote range that we'd seen to the north of us from the uppermost slope of Noshaq. The Soviets wanted to showcase their mountains and were inviting leading climbers from all over the world to the previously closed Pamir range. Months earlier, I had been thrilled when an American Alpine Club board member suggested I apply to be part of the American Pamir team, and then sad when my application was turned down. Having made peace with my disappointment, I now had a summer of protein-folding research planned. I read the cable a third time and urged myself to be rational. Tempted as I was by the adventure of climbing in the mysterious Pamirs, July 13 was less than three weeks away.

The invitation must have been sent at the behest of Heidi Lüdi, a Swiss climber from Rendez-Vous Hautes Montagnes, an international women's climbing club to which I belonged. When she'd heard I hadn't been selected for the American team, Heidi had urged me to join her all-women's team and had lobbied the Soviets—apparently successfully—to extend this invitation to me. Although Heidi and I had never met, she had written that my high-altitude experience would be an asset to her team. But the invitation from the Soviets had come too late.

Putting the telegram down on my lab bench, I tried to go back to work, but my thoughts were in the Pamirs. During a tea break later that afternoon, I told Charles Seiter, a postdoc in my research group, about the Pamir trip.

"You've got to go," he said. "It's the chance of a lifetime."

"I'd have to send money to Russia today," I said. "The bank closes in twenty minutes."

"Then it's not too late," said Charles, grabbing my hand. We ran out of the lab, down the stairs, and jumped on his Harley-Davidson. Zooming into the bank parking lot at 5:00 P.M., we raced inside just as a guard was locking the doors for the weekend.

"I need to send money to Russia right now," I told the teller. She wrinkled her brow, but instead of telling me to come back Monday morning, she figured out how to cable currency to the Soviet Union. As my money went off to the other side of the world, I could scarcely believe that I would soon follow.

Nineteen days later I boarded a plane for Moscow. Once settled in my seat, I thumbed through the official brochure. "Well-known Soviet climber Vitaly Abalakov invites you to conquer the Pamirs," it began. I studied the glossy photo of our ice-capped objective: Peak Lenin, at 23,406 feet the highest in the range. I was pleased to read that Lenin was described as an easy and safe peak, the most frequently climbed 7,000-meter mountain in the world.

I'd packed so quickly that I'd neglected to bring anything to read during the inevitable storm days on the mountain, so when we stopped in Copenhagen I hurried off to buy a book. I reboarded with *The Treasure of the Great Reef* by Arthur C. Clarke, the only book in English in the airport bookstore, tucked under my arm.

As we flew over Eastern Europe, my thoughts turned to my three teammates, whom I would meet for the first time in Moscow. In my purse, I carried welcoming letters from Heidi, but I knew little of my companions' personalities, ambitions, or climbing abilities—nor did they know mine.

At a spartan hotel in Moscow—the Sputnik Hotel—I found my team among the crowd of international climbers: Heidi Lüdi, a sturdy twenty-seven-year-old medical doctor; Eva Isenschmid, twenty-three, a gentle artist and photographer; and Margaret Münkle, fifty-four, an experienced mountaineer. They were all from Switzerland and spoke to me mostly in German, which I could barely understand. Because Heidi and I had corresponded in English, it hadn't occurred to me that they, of course, would talk with each other in German and I would feel left out.

Three days later, all the climbers took a night flight to Osh in Kyrgyzstan, where we were greeted by pink-cheeked children who

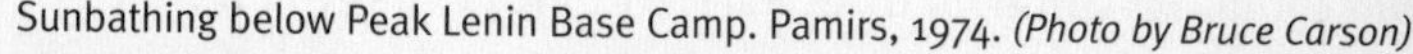

Sunbathing below Peak Lenin Base Camp. Pamirs, 1974. *(Photo by Bruce Carson)*

handed us bouquets of fragrant red roses. Another short flight and a bumpy five-hour truck ride brought us to our Base Camp at 12,300 feet, in a green meadow sprinkled with brilliant wildflowers.

Rows of identical white canvas tents were dwarfed by the icy splendor of Peak Lenin rising directly above. Around us, sheep and yaks grazed in the lush grass, and azure lakes reflected the white mountains. The camp was much more luxurious than I could have ever imagined. The Russians were staging this "sports camp" not just to share their highest mountains with the world, but also to raise hard currency for a Soviet Himalayan expedition. They charged us $750 each, and we got our money's worth: hot showers, volleyball courts and soccer fields, and nightly movies. Electric lightbulbs dangled at the door of each tent. I felt like I had stepped inside the glossy brochure.

Even though I had known the event was going to be international, its scale surprised me. More than 160 climbers from a dozen Western countries enjoyed the amenities of the upscale tent city, while about 60 climbers from Eastern Europe and the Soviet Union resided in a simpler camp across a stream.

Entering the huge mess tent for breakfast the next morning, I walked over to a buffet table laden with smoked salmon, a samovar of hot, sweet Russian tea, and top-quality beluga caviar. I smeared a thick slice of dark Russian bread with sweet butter and then piled it high with caviar. Extravagant, but plentiful in the Pamirs. (Later, when I priced caviar in Moscow, I found I'd been eating about fifty dollars' worth every morning.)

Looking for a place to sit in the tent, jammed with climbers chatting in a complex mixture of languages, I saw my teammates Heidi and Eva at a German-speaking table. They were almost finished, so I headed to an American table and the welcome opportunity to speak English. Shyly, I stood by an empty chair listening to their banter, but no one looked up. I knew some of the American team by reputation, and that there were two young women members, Molly Higgins and Marty Hoey. But half a dozen unfamiliar guys were talking intently.

"Good morning," I said quietly. No one looked up.

"My German's pretty weak, so it's great to hear English," I said, a little louder. Nobody seemed to hear me.

"Do you mind if I join you?" I finally asked, trying to sound matter-of-fact.

They kept talking as though I were invisible. Feeling like an overenthusiastic Saint Bernard, I turned and fled.

I stood amid the roar of voices, holding my bread and caviar and wishing I could disappear. Blessedly, a striking blonde Russian woman waved me over to her table.

Elvira Shatayeva welcomed me in careful English and introduced herself as the leader of a team of experienced Soviet women. She told me that most Russian men didn't think a group consisting solely of women could climb a 7,500-meter peak like Lenin. "We plan to be the first women's team to do it," Elvira said. "Or maybe your group will be."

"Maybe we could all climb together," I suggested, her warm smile putting me instantly at ease.

"This is impossible," Elvira said. She didn't elaborate, but I guessed the Soviets wanted only Russian women on the first all-female ascent of Lenin.

"We cannot climb together, but we can celebrate together," she went on. "We'll have a great party after the climb."

As we walked from the mess tent toward the Soviet camp across the stream, Elvira told me her team's ambitious plan.

"We'll climb up Lenin from the northeast via the Lipkin Ridge, sleep on the summit, and then go down the northwest ridge," she said. "This will be the first ascent of Peak Lenin by a women's team and also the first traverse of the peak by anyone, man or woman."

I was impressed by her confidence and enthusiasm.

"We will succeed." Elvira raised her arm as though in victory. "Soviet women are very strong. We have collective spirit and we work together."

I looked at her radiant face and then noticed behind her the old-fashioned Soviet equipment: canvas tents with button closures,

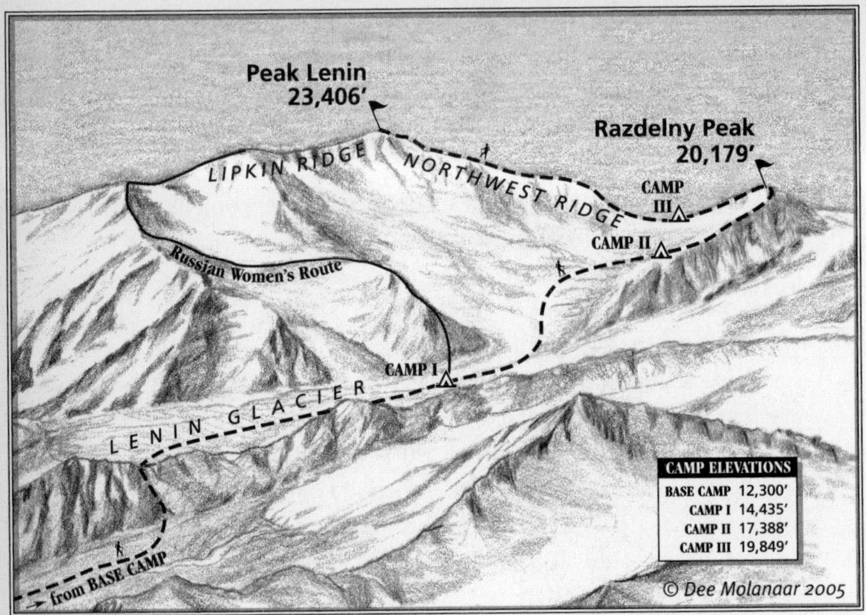

wooden tent poles, boots with nails on the bottoms. The moment was captured as I took a photo of Elvira, her golden-haired profile outlined against the icy summit of Peak Lenin.

After breakfast our team was supposed to begin carrying loads up Peak Lenin. We planned to follow the standard route, stocking three intermediate camps before attempting the summit.

I was eager to start on the mountain, but just as we were about to head up with our loads, a Soviet advisor told us we needed to have our plans approved. Because I'd organized expeditions before, Heidi asked me to discuss our logistics with the advisor. In front of a photo map of Peak Lenin with marked routes, I shifted my weight from one foot to the other while an interpreter translated my words from English to German to Russian and back again. We had to fill out lengthy forms detailing our daily movements. The advisor was dubious of the strength of a party of only four women, and told us to bring extra food. By the time we'd waited in several lines to obtain the rations and repacked our loads, I was inwardly shrieking with impatience.

The Russians were trying to monitor—if not control—our food, equipment, and most aspects of our climb, for our greater safety. I was irritated by this micromanagement and wondered if having others tell us what to do actually increased our security or if it made us less sensitive to the very real dangers of terrain and weather one always faces in the high mountains.

An antlike column of climbers had been coming and going between camps on Lenin for days, methodically moving their supplies up the mountain. At three in the afternoon our little group joined their ranks. My pack was bigger than those of my teammates; unlike European climbers, who tend to carry lighter packs and stay in huts, I was used to backpacking with heavy loads. The air was humid and heavy, and huge thunderheads heralded the daily afternoon storm. An hour after we started, Heidi, Eva, and Margaret stopped to admire a meadow of edelweiss, a flower that had become rare in the European Alps but was still common in the Pamirs. Three Austrians on their way down joined us and began an animated conversation. I waited, the German words washing over me as my heavy pack pushed down on my shoulders. Finally I left and continued up the mountain alone.

I reached Camp I, left my load, and headed back down. An hour later I met Heidi and Eva on their way up.

"Where's Margaret?" I asked.

"She turned back," Heidi said. "Carrying loads at this altitude is too hard for her. She's decided to hike in the meadows with the Italian walkers. Peak Lenin isn't for her."

I was sorry to lose a quarter of our team. As they rose to their feet, I noticed Eva grimacing.

Heidi explained that Eva had stumbled on a loose rock and hurt her rib. As I ran back down toward camp, I worried about the strength of our team. We were now down to three climbers and Eva was injured.

Just above Base Camp, I met the rowdy British team heading up with gigantic rucksacks.

"You look like a strong enough lass to give us a hand with some of this gear." Doug Scott, a renowned British climber, winked at me through his granny glasses. "Want to help carry a load?"

"Sure, I'll help," I said. "A little more conditioning never hurts." And the chance to speak English with these good-looking guys wasn't lost on me.

Doug Scott's strength and energy reminded me of John Hall. Thinking of John made my heart ache. Nevertheless, moving smartly up the slope, I was delighted that I could carry a load for, and keep up with, some of the best mountaineers in the world.

The next morning, I hoped to get an early start up the mountain, but my teammates had other plans. I watched in amazement as they took off their clothes and piled them next to the inflatable washtub, stacked up more clothes from inside the tent, and laid out their shampoo, cream rinse, laundry detergent, and clothespins.

Keeping things tidy had never been a priority in my family. The

Heidi and Eva carrying loads up to Camp III. Pamirs, 1974.

Swiss women, on the other hand, were impeccably clean at all times and seemed to have little tolerance for my casual approach; earlier that morning they had tossed my fragrant socks out of the tent and suggested that I wash them. As I retrieved my socks, mortified, I wondered if Heidi had given any thought to what it would mean to have an American on her team. I certainly hadn't understood what it would be like to live and climb with people who came from a different culture and spoke a different language.

The sight of the meandering stream making its sparkling way through the camp made me think I might as well go with the program and take a bath, too. I followed the stream's leisurely course, looking for seclusion. Each place I selected turned out to be in view of part of the camp. Then I saw a large granite boulder in the distance that would give me privacy.

When I arrived, I found the spot already occupied by Molly Higgins, one of the two women on the American team. Molly, a strong, outgoing woman of twenty-five, stopped washing her long straw-colored hair to wave me over to join her.

I hesitated, because the Americans seemed to be ignoring me.

"Hi, there," Molly said with an affable grin. "This is a great place for a bath and there's room for two."

We chatted for a while and then I blurted out my fear.

"This might sound a little paranoid," I said, "but I feel like you're not supposed to talk to me."

"Totally weird, but true," Molly replied. "Our leader actually told us to avoid you."

"Why would he tell you that?" I asked.

"You weren't chosen for the American team, and you came anyway," Molly said. "That's a pushy thing for a woman to do."

It sounded so ridiculous we had to laugh. At the same time, her explanation helped dull the sting of my rejection. Molly told me that the Americans and the Russians had agreed that no American climbers other than those on the official team would be allowed in the camp. I thought about how the 750 U.S. dollars—a hard currency—I

Heidi bathing in melted snow at Camp II at 17,388 feet. Pamirs, 1974.

had cabled to Russia late that Friday afternoon had been changed into Russian rubles—a soft currency. Apparently by the time the Russians realized their mistake, they couldn't return my money and retract their invitation as the Americans demanded.

I had to smile, thinking of how the intricacies of international finance had brought me to the Pamirs. As we washed, I told Molly the story of the wild motorcycle ride through the streets of Palo Alto to the bank, and how surprised I was to actually be here.

"I'm amazed to be here, too." Molly laughed. A month earlier she had been teaching Outward Bound in the middle of Colorado when a runner arrived to tell her she was on the American Pamir team. "I'd never heard of the Pamirs, but I ran out and here I am. This is my first expedition, my first big mountain."

As we poured water over each other's hair to rinse out the soap, talking all the while, the knot in my stomach relaxed. I found myself wondering why Molly had been invited without having applied. Malinda Chouinard had sent out the names of the dozen experienced women who had requested places on the Pamir team. It was an impressive list. Why had the American Alpine Club turned us all down and invited two relatively inexperienced women who hadn't even sought to go?

When I asked Molly that very question, she wasn't fazed.

"Yeah, both of us are new at expeditions," Molly agreed. "But we work for guys on the team. So they do know us pretty well." That did make sense, as people like to invite climbing partners whom they know and trust.

"And I heard," Molly added, opening her eyes wide, "that several of the strongest women were turned down because they weren't ladylike enough."*

"So for women social skills are more important than climbing skills?" I managed to laugh and ignore the unfairness. Learning the illogic of the team-selection process was a relief in any case. There seemed nothing further to say about the matter, so Molly and I strolled back to camp, chatting.

The Russians had put flagpoles with our national flags in front of the tents, giving the camp the air of a people's summit meeting. To express their feelings about having the Union Jack flying beside their tents, the irreverent British had replaced it with a pair of black lace underpants in the dark of the night. When they climbed out of their tents the next morning, grinning broadly and ready to salute,

---

*Robert Craig, deputy leader of the American expedition, wrote in his book *Storm and Sorrow in the High Pamirs*, ". . . one or two girls from the Northwest had been passed over despite outstanding climbing ability because of their tendency to use gamey language and present a less than 'clean-cut' image."

another British flag was waving proudly, courtesy of the well-prepared Russians.

That afternoon an unseasonable snowstorm covered the camp with heavy powder. Our tents sagged under the weight. "This is very unusual," a Russian advisor told us. "The snow doesn't fall here in the summer."

Waiting out the storm, I went over to the cozy mess tent where the British climbers were drinking vodka and exchanging climbing stories. The Russians, who wanted each team to have a leader, were scandalized by the British, who said, "We're all the leader." The British had decided to name a different leader every day, to the great confusion of the Russians.

As we partied, the Russian women bounded in, jubilant. "We've just made our High Camp at 19,000 feet," Elvira announced. "We're ready for the top."

While the women sang and danced in celebration, I couldn't resist asking Doug Scott why there were no women on his team.

"Most women climbers aren't first-rate," Doug said flatly. "They're so eager to succeed and prove their ability that they don't use good judgment."

"Well, attitudes like that make it tough for women to gain experience," I said. I was telling Doug about our Denali trip and how well the women did climbing Noshaq when two snow-covered Americans burst into the tent. "There's been an avalanche below Krylenko Pass," one of them shouted. "We got out, but there was another team above us. They need help."

A Russian advisor entered the tent. "Do not worry," he said. "Abalakov, the Grand Master of Sport, is confident no one has been killed."

But Abalakov wasn't on Krylenko Pass. I sat with my head in my hands, avalanches thundering through my mind. Finally, an hour later, a Japanese climber came down with the news that a chain of mammoth avalanches had struck nine climbers, but everyone had survived. The Grand Master of Sport had been right after all. Relieved, I went back to my tent, turned off the electric light, and crawled into my sleeping bag.

A few evenings later, I was alone in my tent at Base Camp during another unseasonable snowstorm. As I was dozing off, everything began to vibrate like a dryer filled with tennis shoes. Terrified, I jumped out of the tent, barefooted, wearing only my blue Duofold long underwear. Eugene Gippenreiter, a Russian friend, came by and told me that a powerful earthquake had triggered a series of avalanches.

Molly Higgins, similarly attired in long underwear, ran over and invited me to join her in one of the American tents. "Come on, Arlene. Bruce and I are having a little party in my tent. We've got food and music." I jogged across the snowy meadow and clambered into the candlelit tent. Molly and a slender young man with blond, collar-length hair were staying warm in two fluffy, zipped-together sleeping bags, listening to Beatles tapes.

I had heard of Bruce Carson. At age twenty-three, he was already famous as a pioneer of clean climbing, protecting ascents by fitting small metal "nuts" into cracks in the rock rather than pounding in large metal pitons that damage the rock. I'd heard of his precedent-setting solo ascent of the vertical west face of Sentinel Rock in Yosemite Valley, done without a rock hammer. Bruce had pioneered so many solo hammerless ascents that he was nicknamed "Mr. Solo Ecological."

"Plenty room for one more," Bruce greeted me with a cheerful grin and gestured for me to join him and Molly inside the zipped-together bags. I brushed the snow off my cold feet and put them inside the sleeping bag to warm up. We feasted on cashews, M&M's, and dried apricots, and then began singing along with the Beatles.

"I've got to go write in my journal," Molly said a few minutes later, sliding out of the sleeping bag. "You need this more than I do." She disappeared, leaving me alone with Bruce. I sat up, intending to follow her.

"You don't have to go," Bruce said, giving me a sweet smile. I relaxed back into the warmth of down sleeping bags and companionship.

"This place is lots more comfortable than any other Base Camp I've been at," Bruce said, and proceeded to entertain me with the story of his ascent of the Carstensz Pyramid, a rugged 16,000-foot rock tower rising out of the jungle of Irian Jaya, New Guinea. I listened intently; the Carstensz Pyramid had been on the original Endless Winter itinerary but we hadn't gotten there.

"Wow, I've always wanted to go there," I said, captivated by his modest competence.

"Well, maybe we can go together sometime," Bruce said, moving a little closer to me.

I wasn't sure what to say and pretended I hadn't heard him. "The other place I've always wanted to go is in the Nubra Valley in Ladakh," I said. As I told him of my climbing dreams, it started becoming much warmer in the Pamirs.

Bruce was slowly moving closer to me. His arm was around me. Should I leave? At age twenty-nine, I felt I was of a different generation than Bruce. I decided that since I was six years older, nothing inappropriate could happen and it was okay to stay. Bruce told me about making the first clean ascent of the precipitous Nose route up the sheer face of El Capitan in Yosemite Valley with the legendary Yvon Chouinard. I thought Bruce was a genius of imagination and ethics. He edged even closer. Perhaps a six-year age difference wasn't that great after all. Then I thought of the Americans' disapproval and moved a little away from Bruce.

He reached over, put his arm around me, and kissed me gently. It was getting downright torrid in the Pamirs. I was about to kiss him back, but my mind stopped me. What would people say?

"Gotta go now. Thanks for a great evening," I said, and jumped out of the sleeping bags. I ran barefoot across the snow back to the safety of my solitary tent on the German-speaking side of the camp. That night I dreamt of climbing a mountain covered with marshmallow cream and making love with a handsome stranger.

The next morning we awoke to tragedy. An avalanche triggered by the earthquake had hit four Americans camped at 16,700 feet on the Peak of the Twenty-Ninth Congress. Two of the four were only slightly covered with snow and had dug the third out. But by the time they found Gary Ullin, he was dead. The entire camp was plunged into mourning for Gary, a very well-liked climber. And then the Russians told us that the same earthquake had set off another huge avalanche that scoured the east face of Peak Lenin, where five Estonian climbers had been pioneering a new route. They were also feared dead.

As I lay in my tent crying for the six climbers and in terror of the mountain, Bruce Carson crawled in to comfort me.

"I want to go home," I cried. "These mountains are terrible."

"It's the avalanches that are horrible. Not the mountains," Bruce said, putting his arm around me and stroking my shoulder. "They're just mountains. The Russians say nothing like that avalanche has happened here before. I expect everything will be fine from now on."

"I'm scared," I said, shaking with grief. "So much death. These mountains seem almost evil to me."

"Mountains aren't good or bad," Bruce said, holding me tightly against him. "They're just enormous heaps of rock and ice. They've been here for millions of years and will still be here when we are all forgotten."

As Bruce and I talked about our love of mountains and why we risked our lives to climb them, we took another risk. Bruce kissed me softly and this time I kissed him back. I relaxed into his welcoming arms, feeling protected and safe. I felt like I was choosing warmth and love over cold and isolation. My concerns about what others might think were small compared with my fear and loneliness in the face of the tragedies around us.

After the storms, earthquakes, avalanches, and deaths, couples were forming throughout the camp. As in wartime, people facing the possibility of dying were coming together to affirm life through love.

CHAPTER 16 ▲

My Private Life and Second Place in Violin

Chicago 1959

I pull the flowered wrapping paper off the present Aunt Ruth has given me for my fourteenth birthday. I guess it's a book, and I'm right. On the cover, a teenage girl seated on a small stool is gazing up at the stars and the book's title, My Private Life: A Personal Record for the Teen Years. *I open it and read: "Teen-Time is FUN-time. Here can be recorded by you teeners all the things dear to your heart, the joys and triumphs of the happy teen years of learning and laughter." Not exactly how I'd describe my teen years so far, but I resolve to complete all the pages so I'll have a record, for better or worse.*

I turn to the first entry, "Our Family Tree, the Two Main Branches." I have to leave the page about my father and his parents completely blank. The book and I aren't off to a great start.

Filling out my mom's page is easy until I come to a question about her hobbies and interests. I go upstairs to our room and ask her what she liked to do when she was a teenager.

"Mostly I practiced my violin, a couple hours in the morning before school and lots more hours afterwards," she answers. "I also played tennis and swam. I liked to ice-skate and roller-skate, and basketball was my favorite."

"You played basketball?" I am incredulous.

"I loved it." She smiles. "I was the center on the high school girls' team and we won most of our games."

I try to imagine my mother playing sports. And I'm amazed to learn that there was a girls' basketball team for her back in the thirties while there isn't one for me now in the fifties.

On the next page are questions about my grandparents. So I go downstairs and ask Grandma where she was born. It sounds like she says, "Bileruska near Kiev," but her voice is lost in the blaring of the TV. I ask her where she went to school, but she just stares at the screen.

Grandpa is selling tires at Goldblatts, so I go back upstairs and ask Mom to tell me more about her teen years. She rummages in the closet and hands me a battered shoe box. Flipping through wage records and shopping receipts, I find some yellowed newspaper clippings. I'm surprised to read that Mom won awards for essay writing and academics. The headline of a frayed clipping dated May 1931 declares MISS GERTRUDE ISENBERG WINS 2ND PLACE IN VIOLIN. *Underneath is a picture of Mom wearing a formal dress and pearls. The article says that at age fourteen, my mother won her prize for a moving performance of a Mendelssohn violin concerto. In the photo she looks radiant and confident. It's hard to believe that this beautiful young woman grew into my anxious, hesitant mother.*

"Mom, you look so different here," I say.

She takes the photo from me and shakes her head. "My marriage changed me. It was a big mistake." She stops abruptly, picks up a cigarette and lights it.

"What do you mean?"

"He was six feet two with eyes of blue. A Jewish doctor. I'd had no experience with men. How could I say no? Grandma warned me against him and I didn't listen."

"What did she warn you about?"

"She told me he would be bad for me and he was. Grandma is a good judge of character." She turns the radio to a Beethoven symphony and cranks up the volume. I retreat into The Three Musketeers.

Marrying a tall, handsome Jewish doctor. That's what both my mother and I have been taught should be life's goal. So why did Grandma warn Mom against my dad? What made him so bad?

Although he doesn't come to rescue me, I continue to imagine my father as a good, kind person. When I have to write a term paper for world history, the Holocaust is on the list of possible topics. I know my dad is Jewish and from Germany. I've heard that six million Jews died in the Holocaust, but it's an unmentionable subject in our house, just like my father is. If I learn about the Holocaust, maybe I'll understand more about my dad.

And if I keep asking, I can get Mom to tell me about him. She always gives in when I beg enough.

The Maelstrom ▲

1974

THE FLAGS OF ALL THE COUNTRIES flew at half-mast to honor Gary Ullin and the Estonians. The heavy snowfall—even at Base Camp—had collapsed many tents and turned the dining area into a quagmire. For a day I sat immobile in my tent, considering leaving.

But the sun came out, the air was warm and still, and Heidi and Eva were eager to climb Peak Lenin, its summit shining above us. My teammates' enthusiasm and the momentum in our camp swept me back up the mountain.

When Heidi, Eva, and I tramped out of Base Camp in the early morning, the air was filled with the acrid aroma of wild onions pressed flat from the recent heavy snowfall. Tiny blue stars winked at us from the meadow under our feet. During the next few days, the three of us carried our loads across meadows, rock, and ice, up to Camps II and III. On the warm sunny afternoon of August 3, we reached the 20,179-foot top of Razdelny, a shoulder of Peak Lenin. Passing around a thermos of hot tea, we celebrated our first summit together, toasting in both German and English. At that happy moment, Heidi, Eva, and I were united as a team, our differences forgotten.

That evening, I joined Bruce in his tent at Camp III for a cozy night together before his summit climb. The next day, Bruce and his team headed for the top of Lenin while mine had a rest day. I sat in front of the tent melting snow for drinks and soup while Eva and Heidi relaxed in their tent nearby. I was concerned about Eva. Her rib was better but now she had a bad headache.

"Let's plan tomorrow's climb," I called into their tent. "Would you like something to drink?"

"Not now," Heidi said in a sleepy voice.

"It's important to drink as much as we can today," I said.

"We will eat and drink later when we are ready. Today is a rest day and we're resting," Heidi said, pulling down her tent zipper.

I walked away, thinking about the frostbite of Messner and the Italians on Noshaq and wondering how much I should nag Heidi and Eva about drinking. Back at my tent, I continued melting snow, cooking, drinking, and eating for the remainder of the day.

August 5, 1974, dawned clear and calm. On this glorious morning, I awoke with anticipation tinged with anxiety; today Heidi, Eva, and I would at last attempt the 23,406-foot summit of Peak Lenin. Bruce and his team had reached the top the day before and he told me the climb was straightforward.

He and I shared a quick breakfast in the dark, after which I downed three cups of tea and chugged as much water as I could manage. I filled two bottles with hot lemonade and stashed them along with snacks in pockets I'd sewn on the inside of my down jacket.

A Russian climber came to our tent with a message from the advisors at Base Camp: "A storm is predicted. Do not try for the top."

"Do you know if Elvira and the Russian women are going up?" I asked him, uncertain what we should do.

"Yes, they have left their camp," he said. "They will climb to the top very fast and come down here tonight."

"What do you think about the forecast?" I asked Bruce as I put my lunch into my daypack.

"Not to be taken lightly," he said. "But conditions look perfect right now. And weather forecasts are often wrong. Our track from yesterday is easy to follow. Lots of people are climbing today. If you go light and fast, you'll do fine."

"Elvira's team is climbing in spite of the forecast." I thought for a moment. "I guess we can, too. It would be amazing to meet up with her team on the top of Lenin—a high-altitude women's summit."

As Bruce and I hugged good-bye in the dawn, I didn't want to let go of him. "I'd rather stay here with you," I whispered into his down-jacketed neck.

He nodded and we held each other tightly.

Then, we reluctantly dropped our arms and I crunched across the frozen snow to Heidi and Eva's tent.

"Ready?" I yelled into their tent. "A storm is forecast. We need to leave right away. We have more than three thousand feet, I mean a thousand meters, to go up and down."

"We're just finishing packing extra food and gear." Heidi poked her head out. "In case we can't make it today."

"You're what?"

"We can leave our things in Sepp's tent up high today and then try again another day. Or maybe we can stay up there tonight."

"What about the storm?"

"We will decide when we are up there. If we take the extra things, we won't have to come all the way down," Heidi said.

"I think it's better to go light and fast. If we don't make it today, we can come down and go up again tomorrow," I suggested.

"We will carry food and gear up," Heidi said, and went back inside.

I stood looking at her tent for a few moments, unsure of what to do. There didn't seem to be much point in waiting for Heidi and Eva, who would be going more slowly with heavier loads. I was carrying a minimum and would go up and down as quickly as possible—by myself seemed the only way. So I started slowly up, following a fresh track made by three Japanese who had left before me.

Making a solo attempt on Lenin had certainly not been my plan, but on that calm, sunny morning it seemed a reasonable course of action. A Swiss and an American climber were also climbing alone. Perhaps I could join them or even meet up with the Russian women on the top, as their leader, Elvira, had suggested before.

I glanced at my watch: 6:45 A.M. This was a late start for the 3,500 feet I would need to ascend and descend. I decided to turn back at two in the afternoon whether or not I had reached the summit.

I felt calm and determined as I moved steadily upward, pausing from time to time to sip hot lemonade, nibble a candy bar, and enjoy the spectacular views. The mauves and browns of the steppes of Central Asia funneled out beneath me, punctuated by the icy white peaks of the other high ranges. The only sound was the chomp of my boots on the snow.

Following the trail of fresh footprints and focusing on my breathing, I fell into a pleasant meditative state. I wondered if my family would be proud of my climbing one of the highest peaks in the Soviet Union. Although my grandparents had come from Russia, my grandmother certainly hadn't been happy when I told her I was coming here. I doubted if she or my mother would care whether or not I made the top of Lenin as much as whether I remembered to write or phone them every week. I glanced at my watch and the altimeter in my pocket, which read 11:30 and 22,600 feet. Wow. Almost there and it was still morning. The top was now so close that I thought I might make it by noon.

A blast of wind interrupted my reverie. When I looked up from my feet, the steppes and the surrounding peaks had vanished. I was in a swirling maelstrom of snow. It's a cloud blowing by, I thought. After it passes, the weather will be fine. But the cloud didn't pass, and the wind grew stronger. I could barely stand against the gusts. I fought my way to the top of a steeper section on the ridge. As I moved along a level plateau, perhaps the summit plateau, leaning into the wind, heavy snow began to pelt down. I thought about turning back, but the summit was so close and it wasn't even noon. I continued upward.

By now the wind and snow were eddying around me so thickly that I couldn't see a thing. Damn it, I thought. The summit must be just above, but to go on alone in a storm was too risky. The only sensible direction was down. But where was down? The storm had obliterated the tracks behind me and the whiteout obscured all landmarks. I went down a few hundred feet. The terrain didn't look right. My pulse thumped in my temple and I panted for breath. I retraced my faint trail back up to the plateau, but there were no more footsteps or features of any kind. Everything was flat white.

Were those tracks off to the right? I rushed over and followed some indentations in the snow down from the plateau. I had no idea if they were tracks or if I was going the right way. I hurried back up to the plateau, shaking from cold and fear. All alone in a world of white above 22,000 feet in the middle of Asia. *You may have blown it big this time.*

Then the storm cloud passed by, giving me a glimpse of a faint trail in the distance. I ran over to the footsteps—as much as one can run through snow wearing crampons—and followed them down, praying they would lead me back to camp. A lone figure emerged from another cloud below and headed toward me.

It was Wolfgang Müller, a Swiss climber. *"Wasser, bitte. Wasser,"* he gasped when we met.

I gave him my bottle of lemonade and he tilted his head back and drank for a long time. I took a few deep breaths, relieved to know I was on the right trail and had a companion for the descent.

"Thank you." He handed my empty bottle back to me and turned to continue up the mountain.

"Please, would you come down with me?" I asked.

"The top is close," he said. "I will go up."

"This storm is bad. Please," I begged.

Without answering, he headed up and vanished into a cloud.

I couldn't believe he was still going up into the gale. Should I follow him so I wouldn't be alone? Not a good idea, I thought, and continued down. Suddenly an apparition appeared behind me, running through the clouds above. I waited, and as the figure approached, I recognized Jed Williamson, the American who was also making a solo attempt today. He was coated in snow and looked like an albino yeti.

"Did you get to the top?" I yelled above the roar of the wind.

"Not quite," he yelled back, and we struggled down together.

Farther down, we came upon two figures sitting in a shallow depression in the snow. It was Eva and Anja Vögele, an artist and climber from Bavaria.

"Are you all right?" I asked when we reached them. "Where's Heidi?"

"Searching for Sepp's tent," Anja said. "It's pitched somewhere around here."

"A tent?" I repeated.

"We are sleeping up here tonight," said Anja. "We can reach the summit tomorrow. Stay with us."

"I'm through with Peak Lenin. I'm going down," I said, first in English and then in faltering German. "This storm is dangerous. Please come down with me."

"We will be safe up here, stay with us," Anja repeated. "Heidi will find the tent."

"Heidi will come back. Then we will all go down together," I said.

Eva and Anja didn't seem to hear me; I wasn't sure they understood my poor German mixed with English. My struggle in the storm had left me too tired to try to persuade them further, especially in another language.

Jed was disappearing into the clouds below. I managed to catch up and then followed him down. He was leaning into the wind, breaking a trail for us through deep blowing snow, so I gave him my ski pole to help him stand against the gale. The wind blew me off my feet. I sat, exhausted, wanting nothing more than to sleep in the soft snow. Jed encouraged me to get up and I struggled back to my feet. For what seemed an eternity we continued to fight our way downward through the blizzard.

At last we saw a familiar break in the ridge. Our camp, our blessed camp. As we stepped into the sheltered campsite, there was sudden silence and relief from the storm.

Jed went into his tent and I sprawled down outside mine. Bruce had gone down and I was alone. My hands were icy claws and I couldn't unstrap the frozen crampons from my boots. Just then, a Dutch climber, his red beard caked with ice, came over with a cup of hot chocolate and took the frozen crampons off my boots. I savored the delicious warmth and thawed my hands on the steaming cup. After thanking him I crawled into my frigid tent, too tired to melt snow or eat dinner.

All night long as I tried to sleep I'd hear the wind and startle awake, sick with fear. Of the nine climbers who'd left our camp yesterday morning, only Jed and I had returned. The seven others were up on the ridge above, exposed to the fury of the raging blizzard.

Around seven A.M., Wolfgang and the three Japanese climbers straggled into camp and told us Eva was in trouble. A rescue team headed back up into the tempest with hot drinks, a tent, and sleeping bags.

Several hours later Anja came down with Michel Vincent, one of the rescue team. She told us that Eva was delirious and could not walk. "Eva is talking this morning as though she were a young child with her mother," Anja said. "I think she is not in her right mind."

I prayed the rescuers would find Eva and Heidi and bring them down safely. Interminable hours later, François Valla and Heidi made it back to our camp, saying more help was needed up above. Heidi was okay, but her hands had frozen while she was attempting to bring Eva down. More rescuers headed up into the storm. Shortly thereafter, Peter Lev from America and Sepp Schwankner from Bavaria returned, their heads down.

They recounted how the rescue team had found Heidi and Eva huddled among some rocks, exposed to the force of the blizzard. Eva was delirious, apparently from cerebral edema and hypothermia. She wasn't able to walk or even drink hot tea. The men wrapped Eva in the tent they had brought. Battling drifting snow and winds of eighty miles per hour and greater, they lowered her inert body the eighty-foot length of their rope over and over—more than ten times. As they approached Camp III, the snow was so deep they couldn't move her any farther. Heidi and François headed down for more help. Fearing Eva might be dying, Sepp gave her artificial respiration for twenty minutes. But Eva was dead. Devastated, Sepp and Peter anchored her body in the tent to the slope with an ice axe and struggled back to camp.

I couldn't believe Eva was gone. Sick with grief, I stumbled over to my tent. Tears running down my cheeks, I remembered Eva's joy, her grace. I thought about her sketching the wildflowers of the high meadows, photographing the Kyrgyz nomads in their yurts, sharing her Swiss chocolate on the summit of Razdelny. Could it have been only three days ago?

The storm intensified and the two dozen of us at Camp III found ourselves in our own life-and-death struggle. We didn't know how long we could last here at 20,000 feet. We wanted to go down but the avalanche danger was so great that the Russian advisors radioed from Base Camp not to leave until it stopped snowing and the sun had warmed the slopes for a few days.

Heidi moved into the Bavarian tent with Anja, so I was still alone. Then the Dutch climbers' tent collapsed, and Hans Brüyntjes, who had helped me take off my crampons, asked if he could move into my tent. I agreed, grateful for company.

The Russians came by with brown bread and caviar, a luxurious contrast to our desperate situation. Any Cold War stereotypes I might have harbored about the Soviets being our enemy had been dispelled by their generosity and openness during these terrible days of mourning for Eva and fear for ourselves.

For seemingly endless days and nights the blizzard raged at Camp III and our food dwindled. The Russians stopped visiting. Each small party marooned in its tent struggled alone. Our lives simplified into a battle to keep our tent from collapsing, to melt snow for water, and to cook our remaining bits of food. The intensity of the storm blurred day and night. I became fond of the red-bearded Dutchman who shared my tent.

Hans and I read to each other from *The Treasure of the Great Reef,* transporting ourselves to the blue waters of the Sri Lankan barrier reef. I promised myself that if I got out of the Pamirs alive, next vacation I was going to that warm, safe reef.

When Hans's sleeping bag got wet and froze into a giant ice cube, I had little choice but to share my bag with him. I was awakened in the night by Hans's kisses.

"Who knows if we'll survive this storm," he whispered.

"You know I'm with Bruce," I said, moving away.

"Yes, I know. And I have a girlfriend back in Utrecht," he said. "But we are here now and there may be no tomorrow."

One of my teammates had died. Would I ever see Bruce again? I found it hard to refuse the warmth and intimacy Hans was offering. All of my reservations seemed to come from a place remote from our present world of cold, wind, and snow.

An especially fierce gust of wind ripped our tent on the third day.

We now had no shelter. The wind had eased a little and Hans and I decided to descend to Camp II in spite of the avalanche danger. First, we had to go back uphill several hundred feet along the ridge to the small summit of Razdelny. My large frame pack was like a sail in the wind. I was blown onto my hands and knees, crawled for a while, and then struggled back to my feet. Finally we reached the top of Razdelny, where Heidi, Eva, and I had been so happy so recently. I had no idea of how many days had passed since that tranquil summit, but it seemed another life.

Hans and I headed down through deep snow in a cloud, uncertain of where we were. Suddenly a large crevasse appeared below us, blocking our path. We sank down on the snow. Then we heard a lifesaving yell in the distance. The Russians from Camp II were coming up to show us the way. They found a route around the crevasse, gave us a thermos of sweet hot tea, and led us safely down.

Many of the others who had weathered the storm at Camp III made their way down to Camp II later that afternoon. Conditions were less desperate at this lower elevation, but the storm continued. Hans and I tried to pitch a Dutch tent, but it ballooned away and tore in the gale. There were far too many people for the available space, and I gratefully accepted a cramped place in the doorway of an Austrian tent. It blew down at two thirty in the morning, and I found myself out in the blizzard without my boots or jacket. Someone invited me into a two-person Dutch tent that already had four residents. I was happy to climb in and escape the wind. Hans was the fifth in a small American tent.

The wind stilled in the morning and we crept outside, stiff and exhausted, to see a world transformed. The sun blazed in a dark blue sky. I dug my boots and clothes out of the collapsed tent. Then I lay out on the snow like a lizard absorbing the warmth from the sun and sharing food with the others at Camp II. Initially, I'd felt shy and removed from these elite climbers from Austria, Holland, Russia, and Germany. After weathering the terrible storm and tragedies together, there was a bond of friendship between us. I felt part of a caring international family of climbers.

Around midday Hans continued down with the Dutch team. As I hugged him good-bye, I realized he was two inches shorter than I. I smiled, thinking how I'd never been attracted to shorter men, but of course we'd been horizontal during our courtship.

I lay in the sun, wiggling my toes to warm my still-icy feet. I kept replaying the summit day, trying to change the events so that Eva came down safely. But the ending was always the same. Too distraught to sit still, I decided to head down by myself. Carefully, I

Two climbers make their way down from Camp III during the storm. Pamirs, 1974.

followed the tracks of the Dutch team. The glacier became less steep and I felt the small pleasure of walking on almost level ground, even if it was still ice and snow.

As I stepped with care across some shiny ice above Camp I, a figure heading up waved to me. It was Bruce Carson. I was overjoyed to see him. He ran up and hugged me warmly and led me down to the moraine just above Camp I.

"I have something to tell you," he said. "Let's sit here for a moment."

I sank onto a rubble of dry rocks on ice. After the severity of the last few days, the rock and ice down here at 14,500 feet seemed balmy and soft. Bruce put his arm around me.

"It's about the Russian women's team," he said.

I tensed.

"They died during the storm," he said. "All eight of them."

An explosion of pain and sorrow racked my body, and I was comforted by Bruce's strong arms around me. We mourned and talked together for a long time. Then Bruce had to head back down to his team at Base Camp.

Grief-stricken, I watched a steady stream of climbers silently file by on their way from High Camp down to Base Camp. The mood was solemn, for by now everyone knew of the catastrophe.

Marty Hoey, one of the two young women on the American team, stopped for a few minutes. Before the storm, both she and Molly had reached the summit of Peak Lenin with their parties in good style. Indeed, Molly was the first American to reach the top.

Heidi came by with the Bavarians, her frostbitten hands wrapped in bandages. We hugged sadly and didn't try to talk about Eva. There was nothing to say about our terrible loss.

I was not yet ready to leave the mountain. I sat there on the ice alone with my thoughts. If we had been allowed to go with the Russian women, could we have saved them? Or would we, too, have died? Could I have done more to persuade Eva and the others to come down with me? Why had the worst storm in the recorded history of Peak Lenin blown in just when people were at their most vulnerable?

Today, when no one was climbing, when it didn't matter, the weather was achingly perfect.

What could have happened to the strong Russian women? Most of them had climbed Peak Lenin before. I closed my eyes and saw them in my mind, full of energy and confidence. And now they would never sing or dance or climb again.

I sat piling rocks into a high tower, unable to summon the energy to do anything more complicated than see if I could add one more rock without making the tower collapse. My attention shifted to the edge of the snow. As I watched the slow dripping of water from the melting ice, a vision floated into my mind. It was, of all improbable things, a ribonuclease molecule. And the thawing glacier gave me an idea about how to slow its folding. I would heat up the ribonuclease until it unfolded in the NMR tube. Then I would chill it rapidly and watch it refold at a low temperature where the folding rate is slow. I could take NMR measurements as the protein folded back up. If my experiment worked, I should be able to obtain information about the structure of the partly folded molecule and solve the problem I had been grappling with all year.

The sight of Jock Glidden, Allen Steck, and Chris Wren stumbling down toward me with full packs jerked me back to present reality. They sat down on the rocks, their faces burned and their lips cracked, looking as though they had returned from hell.

And they told me their horrific story. They had been just below the summit of Lenin during the entire storm. When the storm ended, they were so close to the top they decided to head up. Then they saw something red in the distance. It was a woman's jacket, frozen in the snow. Going up, they found a woman's body, stretched out as if sleeping, in a small hollow in the snow. Continuing up they found two more bodies, the remnants of a tent, and then four more bodies. They had come upon the entire team of Russian women, all dead in the snow.

There was nothing to say. Numbly I watched the three Americans head down to Base Camp. Two Russian climbers came over and we talked about Elvira and her team, tears running down our faces.

They told me the women had reached the top of Peak Lenin during the storm late on the afternoon of August 5, the same day I tried for the top. Exhausted and unable to see their descent route in the blizzard, they set up their tents on the summit plateau to wait for a lull in the storm. The weather got worse. After twenty-four hours in that exposed campsite at over 23,000 feet, one of the women became ill. The others tried to carry her down, but could make it only a few hundred yards. They set up their camp on a narrow ledge. They were hit by a hurricane-force wind that destroyed their tents and blew away most of their equipment. They radioed Base Camp, reporting that one woman had died and two others were sick.

With nothing to protect them from the bitter cold, the Russian women had perished one by one on the third day of the storm.

I learned later about Elvira's last radio transmissions: "We are very cold . . . We can't dig a cave . . . there's nothing to dig with. We can't move . . . Our rucksacks were carried away by the wind. It is very sad here where it once was so beautiful."

The Soviet advisors ordered any women who were well enough to come down at once. Elvira responded: "We cannot; we would not leave our comrades after all they have done for us. We are Soviet women. We must stick together, whatever happens."

A few hours later, Elvira came on the radio again: "We have had almost nothing to eat or drink for two days. The three girls are going rapidly . . . What will happen to their children? . . . When will we see the flowers again?"

And finally: "Now we are two. And now we will all die. We are very sorry. We tried, but we could not . . . Please forgive us. We love you. Good-bye."

The radio clicked off and there was only the sound of the wind.

I balanced a last rock on top of the tower I was building and the whole edifice collapsed. Nearly two hundred climbers had attempted to reach the summit of Peak Lenin this past month. Half succeeded. Fifteen died. I couldn't stop thinking of the irony of the warm, sunny weather around me. If it had come only a few days earlier, we would all be down in Base Camp celebrating.

I sat alone, the rubble of my rock tower all around me. The sun shone hot in a bright blue sky. The top of Peak Lenin seemed so close I could reach up and touch it.

Elvira Shatayeva, the leader of the eight Russian women who died trying to traverse Peak Lenin.

CHAPTER 17 ▲

Into Space
1958

On the last day of my freshman year in high school Miss Mitchell asks me to stay after class and talk. "You know, Arlene, you did well in your biology and algebra classes," she says. "In fact, you're doing very well in everything."

"It's amazing," I say. "And I actually like science and math."

"You're third in your class," Miss Mitchell continues. "Have you thought at all about where you might want to go to college?"

Where I might want to go? Outer Mongolia, I think.

"Navy Pier," I say with a shrug of my shoulders. Almost everyone from Bowen High School gets a free college education by living at home and attending the University of Illinois at Navy Pier.

"Well, you might want to think about going to the best school you can."

"But my mom doesn't make enough money to pay for me to go away to college," I say.

"You should be able to get a scholarship," Miss Mitchell says. "But you'll need more than good grades. You've got to be outstanding in other areas, too."

"Like what?" I ask.

"Student government, sports, clubs."

"Thank you sooo much for telling me this," I say. My escape. It could actually happen. I float out of the classroom, already resolving to become so outstanding that there will be no question that I'll get a scholarship.

I plot my self-improvement. I am too shy for student government. I don't write fast enough for the student paper. There are no athletic teams for girls. But by passing some physical tests and swimming a lot, girls at Bowen can earn a school letter in swimming.

That's what I can do. I'll swim and study all the time. I'll go to college far, far away. I'll make a different life for myself.

Tragedy on Trisul ▲

1975

I RETURNED TO MY LAB AT Stanford wanting to resume my research, my friendships, and my old life, but I couldn't stop reliving the deadly days in the Pamirs. I kept asking myself why it had gone so wrong. Some of the events were obviously out of our control: Our gracious Russian hosts' earnest efforts to share their superb mountains with the world had been plagued by an unprecedented series of earthquakes, avalanches, and storms. But some of it came down to human error. Elvira and her team had old-fashioned cotton tents with button closures, not strong enough to withstand the severe storm. Did the coaches, radios, helicopters, not to mention hot showers and electricity, lull us into a false sense of security? Perhaps we had forgotten the fundamental truth that high mountains are never safe or predictable.*

I had lost my passion to climb, or so I thought. Other passions had arisen in the Pamirs. During the storms and avalanches, I had become close to both Hans Brüyntjes and Bruce Carson. Their love had comforted me through death and disaster. But knowing they both had long-term girlfriends back home, I didn't expect to see either of them very soon.

---

*As Vladimir Shatayev, Elvira's husband, later wrote in his moving book, *Degrees of Difficulty*:

> We related to this mountain as to a tame dog whose fangs held no danger to its master. . . . We would ascend to seven kilometers in the sky without doubting the outcome, as assured as if we had been going to the Seventh Heaven in Ostankino [a restaurant on a radio tower in the suburbs of Moscow].
>
> Safe mountains do not exist. Mountains are birds of prey. Sometimes they sleep, full and gratified for many years. It seems to people that they are tame. Everyone, even the most experienced and careful, is lulled by their tranquillity. The precept that there are no safe mountains is gradually erased from memory.
>
> For forty-five years—from the day of the first assault—Peak Lenin had been meek as a lamb. . . . And then the mountain bared its fangs.

As it turned out, my resolve not to climb high was as short-lived as my separation from Bruce and Hans. A few weeks after returning to the cluttered old house in Menlo Park I shared with two Stanford graduate students, I received three letters leading me back to the heights. One, with ornate Polish stamps, came from Wanda Rutkiewicz and Alison Chadwick-Onyszkiewicz in Warsaw; another, with an American commemorative stamp, was from Bruce Carson in Flagstaff, Arizona; and the last, with simple Dutch stamps, was from Hans Brüyntjes in Utrecht, Holland.

Wanda had been climbing nearby Peak Communism when we were on Lenin and had come to our Base Camp by helicopter to talk with me after the tragedies. Despite our sadness, we had continued our discussion, begun in Afghanistan two years earlier, of a joint Polish-American women's expedition to Annapurna. In their letter, Wanda and Alison told me they had applied for a permit to attempt Annapurna I the following spring. They suggested that I invite other American women and begin fund-raising. The Poles, who could not take any hard currency out of their country, would drive across Asia to Nepal in a large truck full of food and equipment for barter. We Americans would contribute dollars.

I appreciated their plan to give women the opportunity to climb to 8,000 meters, but after the deaths in the Pamirs, I was also frightened. I knew climbing big Himalayan peaks was dangerous and that Annapurna I had a worrisome reputation for avalanches.

Bruce's note, recalling the companionship and love we'd shared, contained an enticing invitation. India, he wrote, was opening restricted areas to foreign climbers and he believed we could get permission to visit the Nubra Valley the following summer. Exploring this region had been my dream since peering into the vast wilderness of Ladakh during the Endless Winter.

Bruce already had a team lined up: Phil Trimble, Frank Morgan, and Dan Emmett, with whom he had previously ascended the remote Carstensz Pyramid in New Guinea. Phil, one of the "Three Musketeers," as Bruce nicknamed these Harvard Law School buddies, worked in D.C. and was negotiating with the Indian embassy for our permission. I knew Phil and Frank, having appreciated their comradeship and good humor when we climbed Mt. Waddington together in 1969, and Frank had been a running partner of mine in Berkeley. I hoped these peaks wouldn't be as dangerous as Annapurna, and the idea of exploring this unknown area with sweet Bruce and mutual old friends was hard to resist.

The letter from Hans said he couldn't stop thinking about me since leaving the Pamirs. Unless I wrote him not to come, he'd see me in California in two weeks. The letter was dated two weeks earlier. I gulped. Hans was scheduled to arrive the next day.

Hans, slightly built with long red hair and beard, burst into my life in a whirl of energy. After receiving a doctorate in analytical chemistry, he'd been offered several comfortable corporate jobs but chose to work as a carpenter so he would have more freedom to climb. Bright, capable, and impatient with those who didn't share his intensity, Hans was here to visit me, see America, and climb. Over the next weeks we scaled warm granite cliffs in Yosemite Valley and rocky crags in the High Sierra.

Meanwhile, Bruce, living in Arizona with his girlfriend, continued sending me friendly letters urging me to join his Nubra Valley team. Knowing of my relationship with Hans, Bruce invited him to come, too, and Hans agreed. Climbing with one current and one previous boyfriend seemed potentially awkward, but once again the idea of being part of a family of friends in the mountains outweighed my concern. Our six-person team for the Nubra Valley was complete with the three lawyers, Bruce, Hans, and me.

In November, Hans returned to Holland and I to my research on the problem of how ribonuclease folds. My experiments were done on one of the world's largest nuclear magnetic resonance spectrometers. I was excited to try out the idea for slowing the speed of protein folding that had come to me while piling rocks on the moraine below Peak Lenin. But no one, including my advisor, Buzz Baldwin, believed that it would work. The common wisdom was that it was not possible to obtain NMR spectra showing a partly folded protein molecule.

The Biochemistry Department owned one-third of the Stanford NMR machine, so I was officially entitled to that fraction of the instrument's time. A morning person, I was most productive when first awake, with a precipitous decline in my abilities during the day. I was assigned the 8:00 P.M. to 4:00 A.M. time slot, and the NMR machine was an imperfect prototype still under development. Night after night I struggled to tune the obstinate NMR for my experiment, but I couldn't get it working right. Operating a complex, temperamental machine is not one of my strong points, and doing my research late at night was like playing chess on a miniature set, blindfolded and wearing boxing gloves.

When Hans invited me to come to the Netherlands for Christmas, I gratefully took a break. Based in the twelfth-century house he had renovated in Utrecht, we spent two idyllic weeks touring Holland and skiing in the Alps. I learned to ice-skate on frozen canals and to appreciate herring and onions for breakfast. I was happier than I could remember. The only problem, which I had been studiously ignoring, was the attractive blonde woman who peered out from numerous photos in the house. With feigned disinterest and thumping heart, I finally asked Hans, "Who's the lovely woman in the photo?"

"Nicky, my housemate," he responded. "She's away for two weeks."

"You're still living together?" I said, feeling hurt and jealous. "You told me she wasn't your girlfriend anymore."

"She's my best friend," he said sharply. "Nothing more."

I wanted to believe him, but although I'd been considering staying longer than the two weeks we'd planned, I left as scheduled. I didn't especially want to meet Nicky.

At home I found a letter from Bruce saying that the Indian Mountaineering Federation had changed our objective from the Nubra Valley to Nun Kun (23,410 feet), a complex massif located on the border between the Vale of Kashmir and the high plateau of Ladakh. As I researched the history of Nun Kun, I discovered we would be following in the footsteps of not only Claude Kogan, who had made the first ascent, but also the redoubtable Fanny Bullock Workman.*

Reading about Fanny's adventures so long ago, I was further inspired to move forward on our Polish-American women's climb of Annapurna I. I put out the word in the climbing community that women were wanted for an expedition to an 8,000-meter peak. Irene Beardsley Miller from Palo Alto, Alice Culbert from Vancouver, and Joan Firey from Seattle, three strong and seasoned mountaineers, responded. We now had the nucleus of a team. As it turned out, Annapurna I was closed to expeditions in 1975, because of the nearby CIA camp training Khampa warriors to parachute into Tibet and fight the Chinese.**

Alison and Wanda searched for a different objective and, at the last minute, managed to obtain a coveted permit for Gasherbrum III in Pakistan, the world's highest unclimbed peak at the time. By then, the planning time was too short for us to join them.

In June the Indian Mountaineering Federation told us we couldn't go to Nun Kun and offered us the easier alternative of Trisul (23,406 feet). That was fine with me, as Trisul was known as a straightforward, safe mountain and the approach through the Great Gorge of the Rishi Ganga was wild and scenic. In 1907, Dr. Tom Longstaff made the first ascent, going all the way from 16,000 feet

---

*In 1906 Fanny Bullock Workman and her husband, William Hunter Workman, explored the Nun Kun massif. That same year, at age forty-seven, Fanny climbed to the 22,810-foot summit of nearby Pinnacle Peak. Setting a world altitude record for women that lasted for twenty-three years, Fanny recorded her adventures "for the benefit of women who had not yet ascended to altitudes above 16,000 feet, but are thinking of attempting to do so." Fanny, a staunch supporter of women's suffrage, was photographed at 21,000 feet in the Himalaya with a newspaper headline proclaiming VOTES FOR WOMEN!

**For a fascinating account of this little known endeavor, see *The CIA's Secret War in Tibet* by Kenneth Conboy and James Morrison, University Press of Kansas (March 2002).

Bruce, Hans, and me, shopping for supplies in New Delhi, 1974.

to the top in two long days. For the next twenty-three years, Trisul was the highest peak ever ascended.

Monsoon rains drench the Garhwal in July and August, so we would climb in September just after the rains ended. Meanwhile, all that summer I worked hard to overcome my lack of mechanical ability and tune the complex NMR spectrometer well enough to take my measurements. By shifting the times I ate and exercised, I even managed to modify my biorhythm so I could work late into the night. Finally I was ready to do my experiments and see if I could find direct evidence for an intermediate in protein folding.

One July evening I carefully made up a solution of ribonuclease and transferred it to a small tube that would fit into the NMR probe. After heating the tube to unfold the ribonuclease, I cooled it rapidly to 10 degrees centigrade in an ice-water bath. I then put the tube into the NMR probe which was cooled to the same temperature. As the ribonuclease slowly refolded, I took a series of NMR spectra and was thrilled to see changes that did indeed give me a rough picture of the partly folded ribonuclease molecule. My experiment, which I named "temperature jump NMR," worked just as I had envisioned it while watching the glacier ice melt on Peak Lenin. Although everyone had told me it was impossible to see a spectrum of an intermediate in protein folding, I'd been able to do so. Jean-Renaud Garel, a French postdoc in our group, called it the "Coffee to Champagne" experiment, referring to my moving the ribonuclease rapidly from a hot to a cold environment.* Jubilant at my success, I left for India and Trisul.

On August 11, 1975, our team assembled in New Delhi. Frank Morgan now had a law practice in Djakarta, Indonesia; Phil Trimble was a highly regarded legal advisor at the U.S. State Department in Washington, D.C.; and Dan Emmett was a successful real estate agent in Beverly Hills. And I, who'd always wished for siblings, enjoyed the brotherly teasing and attention of the three lively guys. My relationship with Bruce had made the transition into an easy friendship, but I was confused about Hans. We were once again romantically involved with each other, but I couldn't forget that Hans's "best friend" Nicky was waiting for him in their home back in Utrecht.

In the hot, teeming bazaars of Old Delhi, we found everything we needed for our expedition—from plastic bags to peanuts to Primus stoves. The cases of freeze-dried dinners we'd brought from home seemed boring compared with the local spices, fruits, and vegetables. Our room at the YWCA International Guesthouse began to look like a saltwater taffy factory after an earthquake as the chocolate bars, cheese, and butter we'd accumulated began to melt and run together in the extreme heat.

A few days later we managed to get ourselves and all our gear into three taxis, our snowshoes, ice axes, and arms sticking out the windows. In a wild ride north to the mountains, our taxis swerved amid herds of goats, children skipping to school, gentlemen with briefcases riding bicycles, carts laden with vegetables, and the ubiquitous holy cows. Although our driver never slowed, he managed

---

*Garel, professor of biochemistry at the University Pierre-et-Marie Curie in Paris, wrote me many years later, "It was a 'poker-like' daring experiment, your sense of risk must have been on strike for a few weeks, and most important, it has initiated a very productive (and still active today) area of research."

Buzz Baldwin agreed. "Your work is a landmark experiment that first found physical evidence for intermediates in protein folding," he wrote. "It was groundbreaking and resulted in a whole new field, the study of alpha helix formation by proteins."

Bruce on the way into Trisul in the Garhwal region of India, 1975.

to avoid numerous collisions by inches. Phil, normally imperturbable, stumbled pale and shaken from our cab, glaring at the driver and muttering, "He's going to kill someone before we reach the mountain."

Switching to a more sedate minibus, we continued upward. The road climbed above a roaring river in a nauseating series of hairpin bends, threading its way through the remains of enormous landslides. The tips of snowy peaks began to emerge above the seething monsoon clouds.

Without warning, our bus stopped at a bend in the road and the driver motioned us to get out. A sign saying LATA marked a path to the village where we would hire our porters. Craning our necks, we saw a collection of wood huts perched on a precipitous green hillside directly above. The terrain was so steep that the only place level enough to pitch our tents was on the side of the road.

As we settled ourselves for the night, the mayor of Lata, a small, wiry man with a loud voice, came down and delivered a speech in the manner of a gospel preacher. We were told that for the good, the very survival, of the people of Lata, it was necessary, it was imperative, that we give each porter fifteen rupees and three pounds of rice per day. The fifteen rupees were fine, but supplying that much rice (amounting to eighteen cups of cooked rice per porter per day) would require additional porters to carry the rice for the porters, not to mention porters to carry rice for the porters who carried the rice. . . . So there we sat in the gutter, listening to the mayor's passionate orations for the best part of two days, a face-off between East and West, each with irrefutable logic and justice on its side. Meanwhile, we sorted our gear and cooked our meals on the road, scurrying to safety when an Indian Army jeep roared by.

Western impatience and Eastern ingenuity finally reached a compromise. We would pay the rupees and provide the rice, but there would be no additional human porters to carry the rice. Instead, a herd of goats fitted with small burlap bags joined our team.

And so we left Lata on a rainy morning, the six of us with thirty porters and sixty goats loaded with rice. The first day's trek, directly up 4,500 feet of a sticky mud trail, was exhausting but memorable for the luxuriant, knee-deep wildflowers through which we walked. Picking one of each variety and pressing it inside the book I was reading, I counted more than fifty different types in a rainbow of colors, sizes, and shapes. It was misty and cold as we ascended Dharansi Pass at 13,950 feet and then descended into the tremendous gorge of the Rishi Ganga.

During two days of traversing slippery grass and rocks above the spectacular gorge, the porters chatted and sang, at ease with their massive loads. Following a stream from the Trisul glacier, we reached our Base Camp on a rocky moraine at 15,800 feet. When our unpacking confusion was at its height, large snowflakes began to float down on us. The porters generously shared their hot chapatis and vegetable curry, a welcome change from our cardboard-like freeze-dried dinners.

Scrambling along the "trail" in the Rishi Gorge on the way to Trisul. *(Photo by Hans Brüyntjes)*

The porters had told us stories of "buried treasure" left here by a previous German party. With visions of apple strudel and Swiss chocolate dancing in our heads, we began digging at the former German campsite. A morning's effort yielded eighty-seven used butane fuel cartridges and a mountain of trash. With Bruce's gentle encouragement, we spent the rest of the day burning paper, crushing cans, and packing up loads of garbage to carry out after our climb. The youngest and strongest member of our team, Bruce quietly motivated the rest of us to do the right thing.

Our days evolved into a predictable routine of carrying loads up the mountain and establishing camps. The venerable Nanda Devi, the highest peak in the Indian Himalaya, and her icy companions Kalanka and Changabang dominated our view. Most afternoons, the

Bruce and Hans make their way up Trisul.

mists obscured their majesty, leaving us toiling upward in the heat. The sun reflected off the snow surface and the water droplets in the air, creating an enormous reflecting oven in which we were roasted to perfection.

Heat and altitude were our adversaries as we toiled day after day with thirty- to forty-pound packs. Bruce, who was taking antibiotics for an ear infection and carrying a monster load, nonetheless whistled his way up the mountain. Feeling energetic the morning of my first carry to our High Camp at 20,800 feet, I took a heavier-than-usual load. After an hour I felt as if a giant were pushing down on my chest. I couldn't catch my breath and stopped abruptly, desperate for air. Bruce ran back, pulled off my pack, and sat me down on it.

"Breathe, breathe, relax and breathe," he said, putting his hand on my shoulder. As air filled my lungs, my feeling of suffocation subsided.

"That was scary," I said once I could talk. "I don't know what happened."

"Your lungs probably couldn't expand enough with so much weight pushing down on your shoulders," Bruce said, offering me some chocolate and his water bottle. "I can take a few pounds off you."

"No way," I said, eyeing his massive load. "I'm okay now."

Bruce helped me put my pack on, adjusting the straps so most of the weight was on my hips. We continued up in the scorching heat. To take my mind off my discomfort, I recited a mantra of the name of the dramatic peak above us and synchronized it with my breathing. *Beth-ar-to-li, Beth-ar-to-li.* The syllables became too complicated and I reverted to *one-two-three-four.* Finally inner silence. My pace slowed as I tired.

"Let's unrope," I shouted up to Bruce, not wanting to hold him back. "It's easy from here on. We can each go at our own pace."

"Bad idea," he yelled back. "Always safer to be roped."

Al-ways saf-er to be rop-ed. As we trudged on, his words became my new mantra, a mantra that, if heeded, would have changed everything.

Once at High Camp, Bruce and I dropped our loads, shared a candy bar, and headed back down. On a stretch we were sure was free of crevasses, Bruce coiled up the rope and we ran down next to each other, telling silly stories, singing, and laughing. The clouds cleared and I was filled with joy at being in these magnificent mountains.

The next day we all moved up to High Camp for our summit attempt. When we arrived, Bruce fired up the stove and I began to melt snow for cooking and drinking water. The flame was weak and Hans helped Bruce clean out the wick, but the stove still worked poorly. After dinner, Bruce continued the tedious work of melting snow and was still at it when the rest of us fell asleep.

After the alarm rang at 4:00 A.M., Bruce, who didn't seem his usual cheerful self, confessed that his earache was back and his stomach queasy, but he kept on melting snow while I made oatmeal and cocoa.

At 6:30 A.M. Bruce led off toward the summit with Frank, Phil, and Dan roped behind him. I took a photo of Bruce heading up, his orange parka glowing in the early morning sun.

Hans and I kept the stove going another half hour to melt water for our return, assuming we would catch up with the others by following the trail they broke in the soft snow. We headed up at seven. After an hour, we stopped for a snack and discovered we'd forgotten our lunch. There seemed no hurry, so Hans went back to get it while

Hans Brüyntjes on the way to the summit of Trisul.

I relaxed and contemplated the puffy clouds drifting above the gentle slopes that led to the top. When Hans returned, we continued up, following the broken track on the broad, low-angle summit ridge, enjoying views of Kamet and Nanda Devi.

Without warning a heavy mist swirled in, obscuring the peaks. I felt uneasy, remembering the fierce storm on my summit day a year ago in the Pamirs. But today there was no wind or snow. Pushing away my concerns, I focused on stepping and breathing. Still, I could see only a few feet in any direction, and a vague apprehension floated in the mist around me.

As Hans and I neared the top an hour later, Dan yelled down to us, his words as shocking and unexpected as a gunshot. "Hurry up, we think Bruce is lost!"

From Dan's diary:

> Bruce led all the way to the top . . . sinking in six to eight inches most of the way . . . the first part of the climb was clear and we could see all the mountains. After reaching the top at about 12:30 we thanked Bruce for his Herculean effort in leading the way—in spite of the fact he wasn't feeling too well and was taking tetracycline for an earache. We had a bit to eat, took pictures and movies. . . . The fog cleared locally briefly and we saw southerly of where we were two humps connected to our position, the furthest of which was several hundred yards away and looked close to our height. We were quite sure we were on the highest point. . . . Bruce asked if anyone felt like hiking over to the hump. Phil and Bruce wandered down the hill in that direction. Bruce kept on and Phil returned. The mists returned. About half an hour later the mists cleared momentarily and I saw tracks going down, traversing the first hump below its top and then climbing up to the top of the second, higher hump. The footprints stopped at the near edge of the cornice of the second hump, not returning. The cornice seemed to overlook a steep drop to the left. . . . I called out but no answer. The mists returned.

Hans and I raced to the summit, where we tied Frank into our rope and followed Bruce's footsteps along the ridge into the mist. And then up ahead, we saw Bruce's footsteps leading to a drop-off. Then nothing. I belayed Hans as he crawled to the edge and lowered himself over it. Belay rope around my waist, blood pounding in my head, I waited, hoping against hope that Bruce had fallen only a short way and Hans would find him. The silence was endless as I waited, gripping the rope so tightly my hands felt stiff as claws.

At last the tension on the rope eased. Pulling it back up, I heard Hans swearing in Dutch. "The cornice," he shouted. He reappeared, looking devastated. "He fell through the cornice."

It was too horrifying to believe. In the mist, Bruce had apparently stepped onto a cornice, thinking he was walking on stable snow.

Hans had descended a little way down the nearly vertical slope below the broken cornice. He had seen nothing but the ice and rock face dropping away thousands of feet below him.

"Can't we all go down?" I pleaded. "Bruce could still be alive. He must be." He had to be.

"It's too steep," Hans said. "We could never get to the bottom alive, let alone back up."

"Then we have to find a way around to the base of the face right away," I said. "If Bruce survived the fall, he'll be down there, hurt." Somehow we'd find Bruce. He couldn't be dead.

And then, in the next act of our tragedy, a fierce snowstorm thundered in, obliterating our track from the morning. Just as in the Pamirs, we felt the sting of the wind and couldn't see which way to go. In the distance we spotted a small red flag flapping in the gale. Bruce had left a trail of wands that we could follow down, his last gift to us. Fighting our way from wand to wand, we managed to get back to High Camp.

All that night, as the storm raged, I lay in my tent, willing Bruce to have somehow survived the fall and made his way around the mountain. He would be waiting for us back at Base Camp. *Please, please, let him be there.*

The next day the weather cleared a little and we packed up huge loads and pushed all the way down to Base Camp. I made a deal with

Hans and Frank on the lower summit of Trisul, September 4, 1975.
At left are Bruce Carson's footsteps leading to the broken cornice.

the universe: if Bruce were alive, I'd never complain about anything again. If I could once again see his radiant smile, I'd spend my whole life cleaning up expedition Base Camps all over the world. I felt sick with fear and anguish.

But he wasn't in Base Camp. And the blizzard again grew wild, extinguishing any hope that we could climb the many miles of steep, unknown terrain to the other side of the mountain. Pinned down in our tents for two days, I endlessly replayed the events of the summit day. If Hans and I had left earlier, not forgotten our lunch, warned Bruce not to unrope. If the deadly mist hadn't blown in so he could have seen the treachery of the cornice. But the reality was that no amount of wishing was going to make things different.

Still, I couldn't stop hoping that Bruce would miraculously survive.

Every few hours I struggled out into the storm, calling his name. When I lost sight of our camp I returned, disconsolate.

When the storm finally ended, there was nothing we could do but leave the mountain where we'd lost so much. I carefully folded the bivouac sack Bruce had neatly stitched, and packed away for his parents the gear with which he had done his clean and elegant climbs. Dan hurried out to telegram Bruce's parents and I wrote draft after draft of a letter to them about their amazing son. What could I say? Nothing could ease their terrible loss. Finally I simply told them how happy Bruce had been on Trisul and what a good friend he had been to all of us and to the world. I enclosed a bright yellow wildflower I had pressed during our carefree approach.

Our last night in India, Hans and I wandered aimlessly through the hot, humid Bombay streets lined with thousands of homeless sleeping people.

The reality of despair and death filled my being. Why didn't these people have shelter? Food? Work? Discreetly, I tucked a five-rupee note in the pocket of a ragged child curled up in a doorway. A scrawny old man quickly snatched the money from the child and disappeared. I turned away in misery, nearly tripping on a dead rat floating in a puddle.

I looked up at the skyscrapers thrusting tall into the night sky. Bombay is a center of science and industry. What comfort did the wonders of modern science bring to these thousands of people living in these streets?

I looked with the most compassion at the young sleeping children. Without education, money, or a home, they would live out their lives struggling for food and shelter. I was filled with a compelling need to help make the world a fairer place, a place where these children had a chance. I no longer cared about the intellectual puzzles of theoretical science that had previously impassioned me. I wanted to help reduce human suffering. And maybe my own pain would lessen at the same time.

Hans and I walked all night, and by the time a hint of dawn began to brighten the Bombay sky, I decided to change my life. I would end my academic career as a biophysical chemist and work to help solve the planet's problems. The sleeping shapes in the street were fuzzy and surreal in the early morning light, but I was crystal clear in my resolve.

Back in my Menlo Park home a few days later, I found my copy of *This Is My Voyage,* Tom Longstaff's book about his first ascent of Trisul, and read his description of the mountain's treacherous summit:

> Just beyond us, across a dip in the ridge, was another snow point, sharply corniced on the east, which cut off the view of the south and seemed to me a few feet higher than the one on which we stood: for the summit of Trisul is in form like the two humps of a Bactrian camel.

Longstaff, like Bruce, went from one hump to the next to make sure he had reached the true summit. He cut steps in the ice to the top of the second hump. Because the weather was clear, he saw the dangerous cornice and asked to be belayed.

> Henry stood back to hold the rope in case the cornice gave way as I crawled to the top. I craned over on my belly to look down the astounding southern precipice. Spread below were all the middle hills . . . ; then the foothills; then the plains with rivers winding. To the west all was clear; the whole scarp of the Himalaya so vast that I expected to see the earth rotating before my eyes.

I was overwhelmed with regret and self-reproach. Bruce and I had both thoroughly researched the Nubra Valley and Nun Kun, but when our objective was changed to Trisul at the last moment, our research was superficial. None of us—not I, nor Bruce, nor any of the other team members—had remembered Longstaff's account. His book sat on my library shelf, its critical information far from our thoughts until it was too late.

I went back to my biochemistry lab and told Buzz Baldwin that I was through studying protein folding. I wanted to do practical research that would have a direct positive impact on the world. My work would be dedicated to the memory of Bruce Carson.

With my mind still on the Bombay streets full of sleeping people, global population growth seemed the ideal field. Buzz suggested that I talk to Carl Djerassi, a chemistry professor and pioneer in the development of oral contraceptives, who taught a course at Stanford on birth control and population. Carl welcomed my interest and I joined his class to study solutions to the problem of overpopulation.

As I learned more about the cultural reasons people in developing countries have large families, I began to understand that science and technology alone could not solve this problem.* Nevertheless, having safe, accessible birth control methods for women in developing countries was critical. In Carl Djerassi's class, I learned about a promising injectable contraceptive called Depo-Provera. It was not approved for use in the United States because of concerns that it increased the incidence of cancer. I had an idea about how to study whether this was true. Bruce Ames, a biochemist who had been on my PhD orals committee at Berkeley, had observed that chemicals that cause cancer in animals and in people will also cause mutations, or changes in the DNA, of bacteria. Ames had recently devised a fast, inexpensive bacterial screening test to see whether or not a chemical was likely to be carcinogenic.

I decided to put Depo-Provera through the "Ames test," and stopped by his crowded office in Berkeley to discuss doing a few experiments. Ames, a slender man bursting with enthusiasm and creative energy, listened with increasing attention as I told him that I wanted to change my research from erudite theoretical science to practical work to benefit the world.

"You are welcome to study Depo-Provera in my lab," Ames said.

---

*In India, for example, having male children is essential. Sons care for their parents in their old age and Hindus believe that to have a favorable rebirth, a son must say prayers for them after they die. Given the high child-mortality rate in India, each family wants several sons so that at least one will survive to support the parents in their old age and recite the prayers for a good rebirth. A decrease in child mortality and an increase in economic security of older people would be needed to slow India's population growth.

"But could you also check out some flame retardants?" He explained that a toxic chemical called Tris was being added in large amounts to children's sleepwear to make it flame-retardant. Tens of millions of kids wore treated pajamas, and Ames suspected Tris caused mutations and cancer.

I hesitated, as I knew nothing about chemical flame retardants. But then I remembered what Carleton Coon, a friend who worked for the U.S. State Department, had told me. He said that our planet is being cut asunder by two blades of a scissors. One blade is overpopulation in the developing world. The second blade is abuse and overuse of resources by the developed world. My knowledge of chemistry could perhaps help blunt both blades. And studying a toxic chemical that might endanger children sounded interesting and important.

"It shouldn't take long," Ames said. "You can test the flame retardants at the same time as your contraceptives."

I impulsively nodded my assent, little suspecting that I had just changed the direction of my life and agreed to do something that would affect millions of children. I only wished I could have discussed it all with Bruce.

Bruce Allan Carson, 1951–1975

Chapter 18 ▲

What Goes Up
Chicago 1960–61

My two biggest problems are physics and throwing a ball.

I can't understand what Mr. McClelland, my physics teacher, is talking about. Even though I try really hard, I get a C in his class.

To earn my school letter, in addition to swimming I have to throw a softball into a circle from twenty feet away. I have three chances to pass the test. I fail on my first try. And on my second. Only one chance left.

Desperate, I devise a strategy to deal with both problems at once. With some of Grandpa's black electrical tape I make an outline of Mr. McClelland's round face—complete with mustache—on the side of the garage. Every day after school I throw the ball at him, hard, trying to hit him on the nose. Gradually my arm gets stronger, my aim more sure.

On the day of my third and last attempt at the ball throw, I wake up with a knot in my stomach. In gym class, I go up to the line and visualize the ball sailing right at my physics teacher's face. I take aim, throw, and watch the ball fly straight to the center of the circle. I do it two more times, perfectly. I'm awarded a big B for Bowen High to stitch on my school sweater.

Mr. Rosing is our physics teacher for the second semester. He explains things in a way I understand. My C becomes a B and then an A. For the first time in high school, I get all As. When Miss Mitchell tells me I'm first in my class, I'm on top of the world.

But my time on the summit is brief. Dr. Brachtl, the principal, calls my friend Barbara Gottfried and me into her office and tells us we're not going to be asked to join the National Honor Society because of our "bad attitude."

The top two students in our class, we are stunned.

"Miss Mitchell and Miss Deuter say you giggle and pass notes rather than listen in class," Dr. Brachtl says. "You don't deserve to be in the Honor Society."

Heads bowed, Barbara and I leave her office. How could I have been so dumb? Now I won't get a scholarship to college. I'll be stuck in Chicago my whole life. With my mother and my grandparents. All because I talk to my friends during boring classes.

The next day I realize the world hasn't ended. Even though I'm not in the National Honor Society, I have my school letter. I will go to college far away.

Seduced by Mt. Everest ▲

1976

"HEY, ARLENE, WANT TO CLIMB Mt. Everest?" The half-jesting voice on the phone was that of Phil Trimble. I couldn't believe my ears. Still mourning Bruce Carson's death and trying to refocus my research, I had little interest in high mountains.

"You've got to be kidding," was all I could say.

But Phil was serious. A law school buddy of his worked at the American embassy in Kathmandu and Phil had asked him to check on a permit for us to climb a big mountain in Nepal. He'd cabled back right away that Mt. Everest was available for the next fall because the French had just canceled their expedition. I knew the Nepalese gave only one climbing permit in the spring and one in the fall for each of their big mountains, and Everest was usually booked years in advance.

"We can have the permit if we act fast," Phil continued. "We won't ever have a chance like this again."

"Everest," I gulped. Even for a burnt-out climber, Everest's seductive allure was hard to ignore.

"Frank and Dan are in," Phil said. "With you and Hans, we're a team. What do you say?"

"How can we climb without Bruce?" I finally asked the question that was on my mind.

"We'll dedicate the climb to him," Phil said. "It'll be like he's there on the mountain with us."

I couldn't help thinking this was a rationalization, the kind climbers tell themselves and each other when they continue to climb after a death in the mountains. I told Phil I didn't feel ready for Mt. Everest, and that I doubted if Hans, the three lawyers, and I had the skill to try.

"Think about it. I'll call back tomorrow." With that, Phil hung up.

I sat back down at my lab desk thinking about Mt. Everest. At 29,028 feet, it is by a substantial margin the highest mountain on earth, and an ultimate goal for mountaineers. Only a handful of parties, including one American team in 1963, had reached the summit. And no woman had ever summited it until last spring, when Junko Tabei, from Japan, reached the top. Phantog, a woman from Tibet, made the top just eleven days later. I'd been offered the chance to be the third woman to reach the summit and the first American woman to make the attempt. But so many of my friends had died. How could I consider returning to the high Himalaya?

Then I began to think about what it would be like to try to climb high above everyone and everything to a world of white and wind, snow and sun. The beauty, the companionship, the challenge; where else would they all be so intense? How could I say no to Mt. Everest?

When Phil called back I was still unresolved. "We're nowhere near a strong enough team for Everest, Phil," I said.

"That's true," he said. "With Bruce gone, you and Hans are the only climbers we know. Could you write to Hans? And ask around for some more strong team members."

I agreed to do what Phil asked, suggesting the American Alpine Club meeting in Seattle next month as a possible place to find expedition members—maybe I could even find a woman or two. I had just ventured onto a slippery slope.

The AAC meeting was in full swing when I arrived at the fancy hotel in downtown Seattle. The grand ballroom overflowed with scruffy climbers swapping stories of past adventures and future dreams.

After the opening slide show of a daring first ascent in the Peruvian Andes, we all trooped out for a break. Behind me in line for hot drinks was a guy with a long blond ponytail wearing green hospital scrubs.

"Interesting outfit," I said. "Are you a doc by any chance?"

"Chris Chandler, straight from the ER," he said, shaking my hand firmly. He went on to tell me he'd just come off a twenty-four-hour shift.

"You can function with no sleep?" I asked.

"I'm used to it," he answered. "And besides, I'm hoping to get invited to climb something big. Maybe get to the Himalaya. And it could happen at this meeting. You never know who you'll meet."

Well, you sure are talking to the right person, I thought to myself. The Everest team needed a climbing doctor. Was it fate that one who wanted to go to the Himalaya was behind me in line? I took a deep breath. "Would you be interested in going to Everest this fall?" I asked, wondering what he'd think of such a preposterous invitation from a woman he'd just met.

"Sure," he said with no trace of hesitation. "I'd love to go."

I told him about our possible permit and that we still needed more team members.

"Try Gerry Roach over there," Chris said, pointing to a tall, rugged-looking man drinking coffee at a nearby table. "He and his wife, Barb, are just back from doing a far-out forty-day climb up Mt. Foraker in Alaska."

At the mention of a wife, my interest increased. I went over to Gerry and asked about his and Barb's Alaskan climb. "Sounds like you're ready for Mt. Everest next," I said.

"That's been my dream since I was a kid," he said with a big smile. "I don't know if I'll ever get there, but I've been working toward it my whole life."

"Well, I know of an expedition to Nepal that's looking for members." I felt like Santa Claus. "Would you be interested?"

"Absolutely. What's the mountain?"

"Everest," I said, giving him his gift.

Gerry's grin threatened to split his face.

I told him the story of the three lawyers and the permit, and

Gerry immediately said that he and Barb would take on the enormous job of obtaining and packing all the expedition's food and equipment. Elated at how easy it had been to find a team, I returned home to Berkeley. Hans wrote me a warm letter saying he had decided to try Everest and very much hoped I would, too. Going to Everest was growing ever more alluring. I closed the door on my thoughts of death and began to slide down the slippery slope.

At the same time, my professional life was blossoming. I had become a Bay Area commuter, teaching human biology at Stanford and researching flame-retardant chemicals with Bruce Ames at UC Berkeley.

The basis of my research was a classic drama of good intentions gone wrong. To prevent children's burn injuries, the Consumer Products Safety Commission had established flammability standards for their sleepwear. Manufacturers found that chemical additives—primarily halogenated hydrocarbons, chains of carbon and hydrogen with bromines and chlorines—were the most economical way to meet the standards. Although this family of chemicals was known to be toxic, long-lived, and likely to accumulate in people and the environment, no one thought about how the flame retardant might affect the children who wore the sleepwear.

Tris, the most widely used flame retardant, was layered onto pajamas in amounts of up to 20 percent of the weight of the fabric. When I added a small amount of Tris to petri plates of growing bacteria, the DNA in the bacteria changed, or mutated, showing that the Tris was likely to be cancer-causing. Because it was not permanently attached to the cloth, Bruce thought it could possibly be leached out by a child's saliva, sweat, or urine, and then be absorbed through his or her skin.

As I made my way between Stanford and UC Berkeley on the crowded freeways, my mind alternated between images of the icy splendor of the Himalaya and endangered children wearing toxic pajamas. And as I continued to help plan the expedition, and without making a conscious decision, I slid the rest of the way down the slope and found myself a member of the team. I had been unable to resist the allure of trying to climb the world's highest mountain with my friends.

At the same time, I continued my research. Tris, I found, was broken down into simpler chemicals, or metabolites, once inside a person's body. By reviewing the scientific literature, I made the startling discovery that Tris metabolites were known to cause cancer in animals. Almost all the children in America—including newborn infants—were being exposed to these dangerous chemicals.

Meanwhile, the time had come for my first tenure-track academic job. Since I loved teaching, I applied for faculty jobs at several small liberal arts colleges that focused on undergraduate education. My teaching and research experience made me a strong candidate and I was offered several good jobs. I chose a position as an assistant professor of chemistry at Wellesley College, near Boston. To the surprise of the chemistry faculty, I then requested my first semester off to climb Everest.

Fortunately for me, Wellesley, one of the oldest of the nation's women's colleges, has a long tradition of supporting women's endeavors. The faculty appreciated that being on the Everest team was an opportunity for me, and indeed for American women, and graciously gave me a leave of absence.

Stopping in Chicago on my way back from a visit to Wellesley, I went to see my grandmother, who was in a nursing home on the South Side of Chicago, and my mother, who now lived independently in a pleasant apartment in Hyde Park. My mom and I remained in the roles we'd established long ago: I was the parent and she the child. I helped her open a bank account, buy a television set and an air conditioner, and taught her how to use them. Although my mother consistently proclaimed she had no interests but me, on her coffee table I was happy to find programs from concerts she attended at the nearby University of Chicago and to see her greet others in her apartment building in a familiar manner. I tried, with little success, to stay calm in the face of her monologue on the barrenness of her life without me. I left, assuaging my guilt with the thoughts that her life wasn't as empty as she claimed and that she needed to fill it herself.

About a month after my visit, in March 1976, my eighty-five-year-old grandmother suffered a fatal stroke. I was oddly unaffected by her death. I'm not sure why. Maybe I had built a wall between us from a young age, not wanting to be pulled into her negative world. During my childhood she had been a force to be reckoned with, but she hadn't entered my heart. Many years after Grandma died, I realized she must have had her reasons for taking refuge in soap operas and quiz shows. I finally felt sad that I had never known much about her. She had always been an emotional mystery to me, and her story died with her.

April found our expedition growing. Two Colorado friends of the Roaches joined us: Bob Cormack, an easygoing physicist and adventurer, and Dee Crouch, another high-achieving emergency room doctor. The last person to sign on was Rick Ridgeway, Chris Chandler's good-natured climbing partner.

On most expeditions to very high peaks, there are only enough resources for two team members to try for the top. Our expedition's stated objective was "to put all or as many members of the team as possible on the summit of Everest, demonstrating that an informal and relatively inexpensive assault on the highest mountain in the world can be successful." Putting everyone on top and having a small, low-cost trip were at odds with each other from the beginning.

Typically, ten climbers are needed to carry the oxygen, food, tents, and other equipment needed for two people to make the final ascent. To carry the loads necessary for all eleven members of our team to reach the summit, we hired Sherpas, a tribal people originally from Tibet, who have lived above 10,000 feet for generations. The number of Sherpas kept increasing: ten, twenty, thirty, and then forty, as did the cost of outfitting, feeding, sheltering, and paying them. As our budget skyrocketed, we named our trip the 1976 American Bicentennial Everest Expedition (ABEE) to attract media attention and sponsors. Our fund-raising literature proclaimed: "Two hundred years after the difficult days when our nation was born, we would like to demonstrate that some of our ordinary citizens still possess the

Members of the American Bicentennial Everest Expedition team, 1976. *(From left, standing)* Bob Cormack, Phil Trimble, Rick Ridgeway, Dee Crouch, Barbara Roach, and me. *(In front)* Gerry Roach and Dan Emmett. Frank Morgan, Hans Brüyntjes, Chris Chandler, and Johan Reinhard aren't pictured.

toughness, determination and spirit of adventure that characterized those early days."

The Associated Press and CBS-TV were happy to give us money—on the condition that we take along a journalist and a seven-person film crew. Our small team of friends had ballooned into a complex assemblage of sixty people. The likelihood of finding the close family feeling that attracted me to expeditionary climbing was diminishing.

In late April, most of our team got together for the first time to pack our food and gear in the fragrant Celestial Seasonings warehouse near Gerry and Barbara Roach's house in Boulder, Colorado. When Hans swept in from Holland, I was once again attracted by his charisma and intensity. But then Hans told me he wasn't comfortable with this group of loud Americans, had serious doubts about the team's experience and cohesiveness, and, as the only European, felt very much the outsider. On a more personal level, he confided he was confused about his career and future and how I fit into his life. I worried that Hans was already moody and unhappy, and we were at only five thousand feet.

The logistics for our climb were formidable, and during the previous months Barbara had supervised packing the food for sixty people for three months, a total of 10,000 pounds. Tents, climbing gear, and cooking equipment, packed by Gerry, added another 10,000 pounds, and the two hundred oxygen bottles filled by NASA in Houston totaled 3,400 pounds. We would need more than five hundred porters to carry all our gear to Everest Base Camp. Bob Cormack filmed the packing and then showed his film to us in a speeded-up version. We laughed until we almost cried to see ourselves racing around, packing meals, and filling boxes faster than humanly possible.

Our food and gear were shipped by sea in early June, and on July 28, 1976, we left Los Angeles on Air Siam, bound for the top of the world.

As we arrived in Kathmandu, my excitement ebbed when the film crew made me disembark from the plane three times so they could record it just right. Back from making the thriller *The Eiger Sanction* with Clint Eastwood, the crew aggressively pursued the footage they wanted, and many team members began to resent their typecasting us before the climb had even begun. For the film, they had assigned each of us a personality, and directed us to say things that would follow their script. When Gerry objected to his role as the heroic jock, they told him there was no time for developing complex personalities.

My role was that of an outspoken, abrasive supporter of women's liberation. While I certainly supported equal rights for women, I wanted to express my own beliefs rather than the exaggerated feminist rhetoric scripted for me. I resolved never to appear hurt, angry, or overtly feminist on camera. This was a good idea, but it didn't occur to me that I wouldn't always know when I was being recorded. Indeed, many of us became reluctant to express ourselves when the film crew was around, fearing distorted stereotypes of ourselves on primetime TV. Like the uncertainty principle in physics, which tells us that when you try to look too closely at a particle you alter it in the process, the film crew perturbed our team. In recording us, they changed our dynamic—and often not for the better.

On August 3 our enormous caravan left Kathmandu for the three-week 180-mile trek to Base Camp. By hiking during the monsoon season, we hoped to be in position to head for the top during the short autumn window of calm, stable weather. Trekking through rain and mud, helping the porters across flimsy bridges, and pulling leeches off one another fostered a unique camaraderie. As we made our way across lush, green terraced fields, the monsoon clouds boiling around us, I felt fit and exuberant, leaping from rock to rock, belting out rock-and-roll songs with Chris's girlfriend, a singer who was coming to Base Camp with us. One day I had so much energy I ran up the hill, passing all of our porters and most of the team, singing all the way.

On August 25 we hiked to the legendary Everest Base Camp itself, a relatively level spot on the moraine of the Khumbu Glacier at 17,700 feet. The guys on the team headed up at full speed. Barb and I walked more slowly, stopping to chat with the yak drivers and local villagers. Arriving several hours after the others, I felt healthy and acclimatized, while several climbers who'd raced up were lying flat out in their tents, suffering from altitude headache and nausea. During the next days, they

On a rest day, I hike up 18,000-foot-high Kala Patar for a classic view of Mt. Everest.

recovered while Barb and I worked hard organizing heavy boxes of gear, equipment, and food.

To climb the mountain, we would build a logistical pyramid, carrying food and gear to predetermined campsites on the mountain. Large amounts would be needed at the lower camps to sustain the climbers who moved up to stock the higher camps. Each successive camp would need less gear; ultimately we hoped to put enough tents, sleeping bags, food, and oxygen at Camp VI, our staging area to make summit attempts. All of us would climb the lower part of the mountain many times to make it possible for a few to try for the top.

First we had to blaze a safe path through the formidable icefall of the Khumbu Glacier. This huge labyrinth of giant, unstable ice blocks, towers, and crevasses was historically the most deadly place on the mountain: thirty-four deaths in the course of fifty or so expeditions. Six Sherpas had perished there six years earlier during the filming of *The Man Who Skied Down Everest.* Nevertheless, looking at the icefall, I felt the way I did on the first day of the ski season, surrounded by untouched snow, bursting with enthusiasm to get started.

But when Phil announced the rotation of icefall teams that night at dinner, Barbara and I were the only climbers not included. I told Phil I'd like to help work in the icefall.

"You would?" Phil said with surprise. "I thought you didn't want to lead steep ice."

"True, but I can still put in ladders and help secure the route," I said quietly. I let it go at that, for the cameras were recording our conversation.

Despite my surface equanimity, I was indignant. Several of the guys didn't lead on steep ice, either. All of them, even those who had recently been helpless with headaches and diarrhea, were included on icefall teams, while Barb and I, who had been doing heavy physical work in the very center of the camp, were left out. Barb told me she didn't mind: her throat was hurting, she was accustomed to letting the guys lead, and her goal was to support Gerry while he climbed the

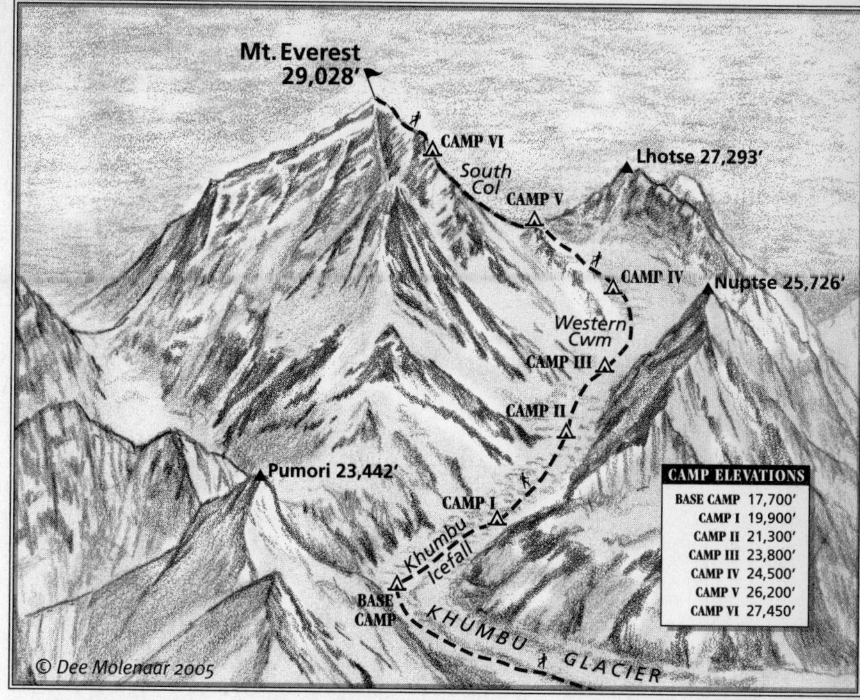

mountain. But I did care that all the men, but none of the women, were asked to work on the icefall.

After the meeting the film's producer walked me back to my tent and expressed his sympathy at the injustice. I told him with some outrage how hurt I felt, a good catharsis for me. Never did I suspect that under his jacket he had a tape recorder, taking down my every word to be broadcast on primetime TV.

The next morning, I made my case to Phil. He listened carefully, apologized, and agreed I could fix aluminum ladders across the crevasses in support of the guys leading through the icefall. I was pleased but couldn't help wondering whether Barb and I would be considered for the summit teams. I'd been part of an in-group of old friends and now I felt myself being edged out. A small crevasse of distrust and misunderstanding began to separate me from the rest of the team.

My first day in the magical Khumbu Icefall began at 3:00 A.M. After forcing a bowl of oatmeal and a mug of cocoa into my unwilling stomach, I stepped out into the cold, strapped on my crampons, and roped up with Johan Reinhard, an American anthropologist living in Nepal who had joined our team as advanced Base Camp translator. Joe, as he preferred to be called, was fluent in Nepali and well acquainted with the local cultures, having spent years living with and documenting the habits of hunting and gathering tribes in western Nepal. He was an intelligent and sympathetic climbing companion who, as his beard grew, reminded me somewhat of John Hall.

Our plan was to fix the ladders for our icefall route in the dark and retreat when the sun began to soften the menacing blocks of ice that hung above us. By the pale light of the stars and in the gray of early morning, Joe and I slowly followed the tracks made by the lead icefall team the day before. We soon encountered a crevasse—about thirty feet wide and one hundred feet deep—where the lead team had bolted together and secured aluminum ladder sections to bridge the chasm. We improved their work by adding handrails and strong anchors. Where the lead team had made their trail around the end of a crevasse under an overhanging serac, we bridged the crevasse with a ladder to make the route shorter and safer.

As the sun rose, I stopped working, looked around, and breathed in the beauty surrounding me. The glacier was a storm-tossed ocean of frozen, brilliant crystal. The icy blue and green walls of the crevasses glowed like sapphires and emeralds when touched by the morning light. This was what I'd come to Everest for.

A huge ice block crashed down just ahead of us, shattering the tranquillity. A surge of fear flowed through me. As the slopes warmed, the mighty glacier began to move. We hurried down, leaving our work to be completed the next day.

For five days, the lead climbers forged a convoluted trail through the icefall. The rest of us followed, placing ladders across crevasses and safeguarding the route. Behind us came the Sherpas, tirelessly ferrying enormous loads of gear and food up the glacier to Camp I at 19,900 feet.

And then it was time to move into Camps I and II in the Western Cwm, the wondrous ice valley leading to Everest's highest slopes. Although the Cwm was a grand abode, life there was not usually comfortable. Temperatures changed rapidly from 90 degrees Fahrenheit

Two porters carrying loads through the ice labyrinth of the Khumbu Glacier on Mt. Everest in the early morning.

in the heat of the day to minus 40 at night. Climbing up to Camp II in the hot sun, we roasted. But when a cloud covered the sun, the temperature plummeted 50 degrees in a matter of minutes.

I had been at Camp II for only a few nights when I began to suffer the consequences of what I now see as my gravest error on the trip, a lapse in judgment involving the simplest and most basic requirement for living organisms: clean water. I had learned during the Endless Winter that rule number one for good health in Asia is *always treat your water.* Accordingly, I had diligently put drops of iodine in all my water in Kathmandu and on the trek. And I had been strong and healthy the whole time. However, once we were above Base Camp, I noticed several people weren't treating their water with iodine, and for some unaccountable reason I stopped treating mine.

One wretched night at Camp II, I had to make six desperate dashes out of the tent. There was no denying it: I had a severe problem. Loud, smelly burps and a gurgling stomach symphony suggested an infection with the protozoan parasite giardia. For a treatment, Chris gave me powerful antibiotics that made me feel even worse.

My gastrointestinal distress decreased my speed and endurance and ended up being perhaps the biggest impediment to my going high on Mt. Everest. Never patient, Hans found my illness and subsequent slow pace intolerable and let me know it in irate outbursts. He apologized years afterward, explaining that it wasn't me that upset him, but rather his uncertainty about his future, his doubts regarding our team's competence, and his fears of the mountain that prompted his unkind words. But all I knew at the time was that my sensitive boyfriend Hans had become angry and accusatory. After numerous failed attempts to talk with him, I found a tent of my own.

And now, sleeping alone at 21,300 feet, I was so cold that I wore all my warmest clothes inside an ultrawarm down sleeping bag, and threw a second thick bag on top. And at least once a night came the inevitable urge that pushed me to leave my warm cocoon for a foray outside. The guys used pee bottles and were mostly spared this ordeal. Once extricated from my sleeping bag and tent, however, I found myself inside a magnificent ice cathedral. In the company of countless stars, I squatted and surveyed the awesome scene around me. The six-thousand-foot walls of Lhotse, the world's fourth-highest peak, and its companion Nuptse were luminous with the white light of the moon. The final eight-thousand-foot wall in the cirque was the massive black pyramid of Everest itself. A tiny speck in this sacred place, I felt a sense of connection to the beauty and goodness of the universe.

A climber crossing a ladder that Bob Cormack and I secured across an enormous crevasse above Camp I on Everest.

My new climbing partner was Bob Cormack; I was relieved to have a ropemate who never blamed or nagged me. Bob had had asthma as a child and was told by a doctor that he would not be able to engage in strenuous exercise. Yet here he was on Everest, proof that for climbs like ours, experience and determination can be as important as physiology.

For two more weeks the Sherpas and team members worked hard, moving loads up to our five camps on the mountain. My mood and hopes fluctuated with the state of my intestines. One day I had such acute diarrhea that all I wanted was to be back home in my warm, cozy bed with a sit-down toilet nearby. The next day I was purged of the evil and back to carrying loads. Trudging steadily up, focusing on the dazzling black summit pyramid with its streaming white plume, I wondered if, somehow, I just might be able to reach that glorious place. A few days

Near Camp II in the Western Cwm of Everest.

later I vomited until I fainted. When I recovered the next day, I asked Bob if we might not make a slow summit attempt together. This cycle of illness and healing, of anguish and elation, repeated itself every few days. No matter how horrid I felt, my ambition to climb high rebounded once I'd regained my health.

By late September, the highest camps were stocked with food and oxygen and the weather was perfect. The time had come for Phil to decide which climbers would be on the summit teams. Although all of us, climbers and Sherpas alike, had worked hard, there was only enough food and oxygen in the High Camps for one or possibly two teams to try for the top. Much of the reward, personal and public, would go to them. Not surprisingly, almost everyone was politicking to be a member of the first summit team.

Phil chose Gerry, whose lifetime dream was to climb Everest; Chris, our climbing doctor; and the Sherpa climber Ang Phurba. The second team was Rick, Hans, and Frank. Bob and I, the slowest climbers, knew we were not under serious consideration. I accepted this, but knowing summit teams can change at the last minute, still cherished a slim hope of getting my chance to try for the top. Indeed, it was up to me if a woman were going to climb Everest on this trip. Barb had lost much of her acclimatization when a severe toothache forced her back to Kathmandu for treatment. After her return, she climbed to Camp II and then generously volunteered to go back down and manage Base Camp as her contribution to the summit attempts.

As Gerry, Chris, and Ang Phurba were preparing to leave, a mail runner arrived from Kathmandu with my absentee ballot for the 1976 presidential election. I cast my vote for Jimmy Carter from 21,000 feet. I also received the galley proofs of a lead article for *Science* magazine that Bruce Ames and I had completed just before my departure. Our paper detailed the dangers of flame-retardant additives in sleepwear and suggested alternatives. Scientific articles traditionally present results without prescriptions for social action, so I was pleased to see that the subtitle we had given our article had been left in: "The main flame retardant in children's pajamas is a mutagen and should not be used."

Bruce's accompanying note told me that additional tests just completed at the National Cancer Institute confirmed that Tris did indeed cause cancer in animals, as we'd predicted.*

Working on the paper gave my spirits a much-needed lift. Although I'd not been included in a summit team, I hoped to carry a load up to Camp V, the South Col, at 26,200 feet. From there, at just 8,000 meters, I could fulfill my dream of looking across to the high arid mountains and brown hills of Tibet. The South Col would be my personal summit.

The night before the first summit team was scheduled to head upward from Camp II, Gerry was stricken with diarrhea and decided to wait for the second team. Another climber had to be chosen. To everyone's surprise, Phil picked Bob Cormack. I was delighted that

---

*This paper also helped bring to the attention of the scientific community the fact that most domestic fires are caused by smoldering cigarettes, and that cigarettes continue to burn because of chemicals added by the tobacco companies.

Our call for self-extinguishing cigarettes was not heeded until more than a quarter of a century later, when in 2003 the state of New York mandated them. If such a law were to be enacted by all states, it could prevent thousands of deaths and injuries from fires started by smoldering cigarettes, and save many millions of dollars as well.

Bob, the least competitive climber on the team, would have this opportunity. Feeling strong and healthy, I offered to carry a load up to Camp V at the same time. At first Phil thought this was a good idea, but he then changed his mind, worrying that if I had a problem during the carry, it would jeopardize the summit attempt. I was disappointed, and wondered if the decision would have been the same if I were a man. Many of the guys who were carrying to the South Col, including Bob, had been battling diarrhea, too. The crevasse between the team and me grew wider.

Phil did agree that Joe and I could carry loads up to Camp IV at 24,500 feet. The next day we moved steadily upward, higher than I'd ever been. I entered a place of brilliant white and deep blue, a stark world dotted with the gaudy specks of orange, purple, and green of our climbing team. The higher we went, the more spectacular and unearthly the views became.

As we approached Camp IV, I was in a meditative state and found myself thinking about my grandfather. I wished he—the person in my family who most encouraged me to aim high, and not let my being a woman interfere with doing what I wanted—could see his granddaughter heading toward the top of the world. I didn't think I would have gotten to Everest without his belief in me.

Finally we were at Camp IV, three small tents pitched on platforms chopped into the ice clinging to the steep Lhotse face. I wanted to go higher, but this was the point at which the summit team needed oxygen to aid their sleep at night, and all the oxygen at the camp was for their use. Joe and I would have to head back down to Camp II.

But first I needed to say

When I reach 24,500 feet, my high point on Mt. Everest, I fly a flag recalling my work on the brominated flame retardant Tris.

my farewell to the high slopes of Everest, and to my hopes of reaching the top. Alone, I headed up a few hundred feet above Camp IV and looked longingly up at the South Col, a mere 1,500 feet above me, outlined against the navy blue sky. I could visualize myself at the Col, looking down on the brown hills of Tibet, but I knew it was not to be. I consoled myself with the thought that 24,700 feet was an altitude record for me. For these brief moments, this beauteous place was my private domain. But soon, whether or not any of us made the summit, we would all leave. The wind would scour away our footprints, the snow would fall, and there would be no trace of our ever having been on Mt. Everest.

At 4:15 P.M. on October 8, 1976, Bob Cormack and Chris Chandler reached the place where the earth fell away beneath them on every side—the summit of Mt. Everest.

Back in Camp II two days later, Bob told me the story of their ascent and I tape-recorded his words for the book I was planning to write about our expedition.

"We started up that spectacular ridge about 7:15 A.M., really a bit late. Ang Phurba went about fifty feet and collapsed. His oxygen regulator had frozen up. We couldn't make it work and A.P. had to give up and go down. Chris and I finally started up at 8:15 A.M., really late.

"At 1:15 P.M., we went over the South Summit of Everest and were hit by the wind. If this had been a climb in Colorado, I would have turned back. I thought about all the people who had worked for many months . . . To make progress, I would kneel down, breathe oxygen, then take ten or fifteen quick steps, and then kneel down. I was near the end of my ability. At 3:55 P.M., we were fifty yards away, so close we couldn't turn back. At about 4:15, I stepped up onto the top, thinking, well, here it is, the summit of Everest. Who would have ever guessed that I'd ever be here? I was surprised it was so small, just a rounded ridge, not a dome at all. The wind was so bad you couldn't take your mittens off more than a few seconds. Chris wanted to shoot some pictures, but our survival was at stake and I said we better get the hell out of here.

"We did start down in half an hour. The wind picked up even more and going down was scary. It took an hour and a quarter to get back to the South Summit. It was dark and the next hour and a half to Camp VI seemed like an eternity. I was really wiped, more tired than I'd ever been in my life. Next morning there was no more oxygen. I just stumbled and fell most of the way down the ridge.

"Now that I'm back down here at Camp II, I can think about it reasonably. I didn't enjoy the mountain. It was the worst pummeling my body has ever taken. I was afraid the whole time. I feel like we just squeaked through a closing door and escaped with our lives."

I fed Bob hot drinks, and then heard on the radio that the second summit team of Gerry, Rick, and Hans had headed up to the South Col that morning. But as they approached the Col they met the Sherpas streaming down. The Sherpas said they were too tired to carry any more loads and that the weather was bad and conditions dangerous. No amount of talking or offers of extra payment could convince them to stay. Then a storm blasted in, blowing away any remaining momentum for a second ascent. Gerry, Rick, and Hans, all bitterly disappointed, had no choice but to come down.

And then it was time to clean up our camp and leave Everest to the winter winds. As we retraced our steps back toward Kathmandu, I was invigorated by the thick air and the vibrant grass and trees. The fall weather was crisp and glorious, the peaks resplendent around us. In spite of my own disappointments, I bounced down the trail, fueled by the rush of oxygen.

Hans's unhappiness about not getting to try for the top only deepened his depression about the lack of direction in his life. While hiking down we had inconclusive long talks about his future. No matter what I suggested, he responded negatively. The world around us was so beautiful, I finally ran out of patience with his grumbling.

"Why don't you just grow up?" I said abruptly.

"You're absolutely right," he said, picking up the pace.

When we reached our camp, Hans walked over to the radio and asked if he could call the Netherlands. He returned a few minutes later, looking happier than I'd seen him for months. He told me that he had done what I had suggested. He was going to become an adult; he had just asked Nicky to marry him and she had accepted. They would be married as soon as he got home.

Needless to say, this wasn't exactly what I'd had in mind. Hans's sudden engagement left me feeling surprised and abandoned. But I rationalized that I didn't want to be with anyone that changeable anyway. And if Hans was looking for a stay-at-home wife, I wasn't ready for that role. Before we left the foothills of the Himalaya, I was thinking about returning.

Back in Kathmandu, I discussed various possible climbs with the Ministry of Tourism, which gives the necessary permits. Annapurna I, which Wanda had suggested as an objective when we met in Afghanistan in 1972, was available for fall 1978. Considerably lower than Everest and famous as the first 8,000-meter peak ever ascended, it seemed a worthy objective. I applied for permission for a team of women to attempt it.

I had intended to write a book about our Everest expedition but I had little desire to spend months or years reliving this trip. I handed the project and my tape-recorded interviews over to Rick Ridgeway, whose book about the climb, *The Boldest Dream,* was published by Harcourt Brace Jovanovich in 1979. I was gratified to read an excerpt of a letter from Hans that Rick chose to include in his book:

> All of us to a certain extent had male chauvinist traits and Arlene was the victim. She more than Barb wanted to be a climbing member of the group and contribute her share in the route making. Yet she never got a fair chance to do it. In the Icefall the chosen lead climbers, including myself, were too busy gaining merit for themselves to let a woman interfere with that and the fact that Arlene was a little slower than the men was gratefully accepted as the real reason.
>
> After that she fell sick, providing the godsent confirmation to the argument that women are too weak, too slow. . . . Finally she recovered but everyone hesitated to believe it, mainly because she became an unwanted competitor in the silent struggle for a place on the summit teams.

> Looking back on it, I fail to understand it. Even Gerry got sick and it was accepted as normal. . . . The conclusion can only be that a woman has to be stronger, better, and more experienced, just to be accepted as an equal.

When I read Hans's letter now, I appreciate his affirmation that as a woman, I had indeed been treated differently. But I had probably overreacted, and added fuel to the subtle sexism that was not unusual in the 1970s. Consequently, and sadly for me, by the end of the trip, I felt far outside my previous circle of friends.

By the usual measures, our Everest expedition was a success. In only the second American attempt on the mountain, two of our team reached the top. I had set an altitude record for American women. We were all alive and in good health, having survived the Khumbu Icefall, the Western Cwm, and the extremes of nature, not to mention the film crew and our own personal avalanches. A team of amateurs, we'd succeeded in a very tough climb. Nonetheless, the trip was not an emotional high point in my life. Looking back with the benefits of maturity and hindsight, I understand that I was caught up in an unfortunate confluence of poor decisions and bad luck, ranging from my deteriorating romance to the macho film crew to my failure to treat my water. And probably coloring my whole experience was the fact that the climb took place only a year after Bruce Carson fell to his death.

Still, the expedition taught me how to organize and implement a Himalayan expedition. Using these skills, I planned to return to Nepal with a team of women to attempt Annapurna I, one of the world's highest and toughest mountains. Finally, I hoped, I would have the opportunity to climb high enough to see the brown hills of Tibet.

Chapter 19 ▲

Thanks to Sputnik
Chicago 1961–62

I love chemistry. I love the order of the periodic table, the way the elements are arranged, each with its own personality. There's lithium, sodium, and potassium, excitable and reactive like Grandma. Carbon, silicon, and lead are stable and well balanced, like Grandpa. Helium, oxygen, and fluorine are volatile like my mother, suspended in the ether. But there the resemblance ends. Chemicals are always the same, constant and predictable, not at all like my mercurial family.

Sy Vogler, my aunt Ruth's husband and my favorite uncle, is a chemist at Argonne National Laboratory. I love talking with him about the magic of combining elements to make molecules. I don't want to become a kindergarten teacher anymore; my ambition is to become a chemist like Uncle Sy.

My mother supports my new career. As a chemist, I will be independent and won't need a husband to take care of me. Grandpa gives his approval; chemists earn lots of money. Grandma's worried; what nice Jewish boy would want to marry a chemist?

I study harder than ever for my College Boards. On the day of the exams, I feel like I've been preparing my whole life. I've tried to memorize the dictionary to increase my vocabulary. I've read books on the psychology of test-taking. My hands are sweating and I'm breathing hard as I walk into the austere classroom where the exam is to be given. I sit down at a battered wooden desk, grasp my pencil, and wait anxiously for the monitor to tell us to begin. I imagine myself scoring perfectly as I work my way through the questions.

In the weeks after the test, my stomach flutters whenever I think about it. My life depends on the outcome. When the envelope comes, I open it with shaking hands. My scores are high enough for a scholarship. I'm saved.

In April, I'm ecstatic to receive fat envelopes from all the colleges to which I applied. I choose Reed College, a small liberal arts school in Portland, Oregon, because it gives me the best scholarship—$2,000 per year, which covers full tuition, room, and board. Also, Portland is two thousand miles from Chicago.

The night I send in my letter of acceptance to Reed, I go outside and sit on the porch. I look up at the night sky and offer a special thanks to my old friend Sputnik.

Annapurna: Women in High Places ▲

1978

ONE CLOUDY SEPTEMBER afternoon at Base Camp, I set my camera on a tripod and managed to get everyone lined up for the group photo I'd been trying to take ever since we arrived in Nepal. I looked through the viewfinder at twelve smiling faces, the 1978 American Women's Himalayan Expedition, the first women and the first Americans ever to attempt Annapurna I.

Standing in the back row on the far left was Margi Rusmore, our youngest member, a twenty-year-old geology student at UC Santa Cruz who climbed Denali when she was seventeen. Next to her wearing a British Airways shirt was Alison Chadwick-Onyszkiewicz, thirty-six, the British artist and veteran climber with whom I had first discussed this trip in Afghanistan in 1972. Three years earlier Alison and Wanda Rutkiewicz had made the first ascent of Gasherbrum III in Pakistan, the world's highest unclimbed peak. On Alison's left was our Base Camp manager, Christy Tews, a thirty-eight-year-old Kansas City housewife who until recently had never even seen a mountain. Then came Piro Kramar, forty. Piro, an ophthalmologist from Seattle, was our team physician, reserved and strong. In the center, beaming, was Irene Beardsley Miller, a physicist at IBM and the mother of two teenage daughters. She had gone to Nepal in 1961 with her husband, who was a member of Sir Edmund Hillary's Makalu expedition. Able and eager to climb, Irene had been forbidden by

Members of the American Women's Himalayan Expedition to Annapurna I at Base Camp, 1978. *(From left, back)* Margi Rusmore, Alison Chadwick-Onyszkiewicz, Christy Tews, Piro Kramar, Irene Beardsley Miller, Joan Firey, Annie Whitehouse, and Marie Ashton. *(Front)* Dyanna Taylor, Vera Watson, Vera Komarkova, Liz Klobusicky, and me. *(Copyright 1979 by the National Geographic Society)*

Hillary to set foot on the mountain; now forty-three, she was worried she might be too old to go high on Annapurna. Next, with a scarf around her neck, was Joan Firey, forty-nine, an artist, physical therapist, and veteran of tough ascents in the North Cascades. Arm in arm with Joan was Annie Whitehouse, a twenty-one-year-old nursing student and climbing partner of Margi's. Annie's other arm was around Marie Ashton, the talented and sensitive sound engineer for the documentary film we were making about the ascent.

Kneeling in front on the left was Dyanna Taylor, our brilliant and insightful cinematographer. Then came the two Veras. Vera Watson, forty-four, was a computer scientist at IBM. Determined and fearless, she had been the first woman to make a solo ascent of Argentina's Aconcagua, the highest mountain in the Western

Hemisphere. At thirty-five, Vera Komarkova was a climbing powerhouse who had emigrated from Czechoslovakia and earned a PhD in high-altitude plant ecology. Next to Vera K. was Liz Klobusicky, thirty-three, a skilled technical climber who taught English in Germany.

I made sure the camera was focused and then asked Lobsang, our head Sherpa, to push the release. I ran over and knelt next to Liz, taking my place in this family of women, a family with whom I would struggle, celebrate, and mourn.

A year and a half earlier I had begun teaching at Wellesley College, trying to instill a love of chemistry—not always successfully—in my young students. At the same time, our Annapurna expedition was moving forward. Vera W., Irene, Joan, and I had received tentative permission from the Nepalese to attempt Annapurna I, but to get our official permit we had to be endorsed by the American Alpine Club (AAC). Unexpectedly, and for no stated reason, the AAC stalled on taking action. A friend on the AAC board told me that the committee felt they had to be much more careful when approving a team of women, fearing bad publicity if things didn't go well. When the AAC finally did notify us, it was to say they would endorse our team if we had a different leader.

I was stunned. The AAC didn't dictate to men's expeditions who their leader should be. Adding to the hurt and insult, my friend on the board told me the committee was against me because of my pushiness in going to the Pamirs when the Americans hadn't invited me. Rather than just shrugging the incident off as unfair and ridiculous, I reverted to my old habit of blaming myself, feeling that there must be something wrong with me if the climbing establishment wouldn't accept me. To my relief, Vera W. took on the job of convincing the AAC to endorse us with me as leader. After much lobbying she finally prevailed, and we moved forward to selecting the rest of our team, organizing our equipment and food, and fund-raising.

All the preparations for the climb were centered in the West and I felt isolated in my basement apartment in Wellesley. I was home alone feeling very lonely one Saturday in April when the phone rang at 10:00 P.M. I answered it, hoping it was a friend, any friend.

It was Bruce Ames, calling from Berkeley to tell me that our scientific work had paid off. The government had banned Tris from use as a flame retardant in children's sleepwear. I was thrilled, but then outraged when Bruce went on to say that many pajama manufacturers were instead using Fyrol, a chlorinated chemical that had virtually the same structure—and was likely to be as toxic and long-lived—as Tris.

"There's still a lot of work to do on flame retardants and you're the best person to do it," Bruce said. "How about coming back to Berkeley? There's a position for you here in the Biochem Department."

I smiled to myself, thinking no one but Bruce Ames would phone me at home at ten o'clock on a Saturday night to offer me a job. I was tempted. In Berkeley, I could continue my research and also be closer to high mountains and the Annapurna team. It seemed a good omen that Bruce called just when I needed a friend. I accepted his offer and told Wellesley that I wouldn't be coming back the next year. Once again, after a short stay on the East Coast, I threw in my lot with California.

Back in Berkeley that summer, I decided to see if Tris-treated sleepwear had indeed been removed from the stores as the law required. Feeling like a superspy, I rode my bike to JCPenney and bought two dozen sets of children's pajamas in a wide range of brands and sizes. "What a wonderful large family you have!" the saleslady gushed.

Taking a rucksack full of pajamas back to my lab, I extracted the flame retardant from the fabric and looked at the solution in an NMR spectrometer. The signal clearly indicated the presence of Tris or Fyrol in half the pajamas I'd bought. Friends around the country sent me samples of sleepwear from their local stores, and the results were the same. It appeared that half the pajamas being sold after the ban still contained Tris or Fyrol. Bruce and I wrote another paper announcing our findings, and retailers finally removed the offending sleepwear from their shelves.

We then decided to study whether the toxic chemical was absorbed through children's skin and, if so, how long it stayed in their bodies. We collected urine samples from ten children, eight of whom had worn Tris-treated sleepwear prior to the ban three months earlier; the other two had not. We found breakdown products from Tris in the urine of the eight children and no Tris metabolites in the urine of the other two.

Bruce and I published another paper in *Science* magazine showing that both Tris and Fyrol not only were absorbed but also could be found in the bodies of children six months after they had stopped wearing Tris-treated pajamas. Although these chemicals were finally off the market, millions of pounds of treated sleepwear still existed. The manufacturers proceeded to export them to the developing world.

Meanwhile, Vera W., Irene, Joan, and I worked on assembling the rest of our Annapurna team, inviting candidates with the requisite high-altitude experience to climb with us in the Sierra Nevada. After a few days together, we could see whether the applicant was a good fit. We were surprised to discover that some outstanding Yosemite technical rock climbers lacked experience cooperating with a team and dealing with adverse weather conditions—the necessary skills for expeditionary mountaineering.

Our selection process worked well, and by late fall of 1977, we had a strong ten-member team. Getting in condition for the climb was vital; whatever time we could spare from organizing, we spent climbing, running, lifting weights, and carrying backpacks filled with books and gallon water bottles up steep hills.

Earlier, Irene had suggested that she and I try to run a marathon. To my amazement, not only was I able to complete the twenty-six-mile course, I actually enjoyed it. I then had the privilege of running in the first-ever U.S. women's marathon, held in Minneapolis in October 1977. About two hundred women distance runners, many of whom had been excluded from previous marathons because of their gender, took part in this festive race.

Running in the first-ever U.S. women's marathon, a joyful event held in Minneapolis in October 1977.

At the same time, we were trying to raise the $80,000 I'd estimated the trip would cost. I recruited everyone I met to be an expedition volunteer, and together we brainstormed fund-raising possibilities. From banquets, balls, and Bay cruises to tournaments, treks, and T-shirts, the moneymaking schemes were debated in my living room.

Finally we decided to sell T-shirts emblazoned with the daring slogan A WOMAN'S PLACE IS ON TOP . . . ANNAPURNA, dreamed up by my creative Reed College roommate Sylvia Paull. Whether to support the trip or because they liked the slogan, everyone wanted one of those shirts. Fifteen thousand T-shirts later, we'd raised much of our $80,000, with advances from an article for *National Geographic,* a book contract, and a documentary film contributing the rest.

Christy Tews, who had recently arrived from Kansas looking for an adventure, parked her camper in front of my house and became a full-time expedition volunteer, taking charge of the unruly T-shirt business. And in a contribution to my personal life, Christy introduced me to John Percival, a tall, contemplative carpenter who became my boyfriend and a source of support, stability, and love during this hectic time.

In an unusual step for an expedition, we decided to meet with a psychologist to discuss group dynamics. Many of us were skeptical about taking time from our preparations for so nebulous an objective, but as it turned out, this meeting was critical to our success. We'd heard predictions that a group of women would crack under the pressure of a risky high-altitude endeavor. During our first session with the psychologist, she asked how we planned to handle disagreements. With her guidance, we made a commitment to talk out our differences and actively work to avoid the miscommunication, conflict, and resentment that can trouble a Himalayan expedition. And then she asked what sort of leader the team wanted.

"Someone who's strong and decisive, who will make firm decisions and stick to them," said Vera W.

"But we all want to be part of the decision-making process," Alison said, looking pointedly at me. "It will be good for the group and will lighten your burden."

I couldn't help wondering what it meant to be the strong leader of twelve tough-minded women, each of whom wanted to contribute to every decision. A bit flummoxed about how I was going to meet this challenge, I was nevertheless grateful to have it articulated before we were on the mountain. I already knew that my personal challenge would be to make clear decisions—and then stick to them. As a scientist, I tended to see all sides of an issue, and to replay decisions even after a choice was made. I knew this style of leadership wouldn't be practical or safe during the inevitable crises of a high-altitude expedition. With an avalanche approaching, I would need to tell the team which way to go; deliberation and democratic decision making could be fatal. I resolved that on Annapurna, I would be a model of decisiveness.

In addition to balancing the needs, skills, and personalities of the team members, I had to direct a small army of staff. Nepal has few roads, and our approach to Annapurna included bridges and steep slopes that could not be crossed by pack animals. Our ten thousand pounds of food and equipment would be carried into Base Camp on the backs of two hundred porters.

I thought climbing and load carrying high on the mountain itself would be faster and safer if we had Sherpa porters to help, but Alison and Liz felt that men working for us on the mountain detracted from this being a women's expedition. After heated discussions, we compromised by hiring five Sherpas as high-altitude porters—in contrast to the forty we'd employed on Everest in 1976.

Our plan to hire these Sherpa women as high-altitude porters on Annapurna I didn't succeed.

On August 16, 1978, our high-spirited caravan of porters, Sherpas, kitchen staff, friends, and climbers finally left Pokhara, Nepal. Pelted by the monsoon rains, we trudged up muddy paths to chilly, windswept passes; back down to green terraced rice fields; then up once more. Twelve days later, we reached Annapurna North Base Camp at 15,000 feet. We planned to set up and stock five camps along the flanks of Annapurna to make it possible to try for the top. That day seemed a very long way off as we gazed at the summit looming more than two and a half vertical miles above.

Our first morning at Base Camp, Liz and Alison set off and in a few hours established the route to Camp I at 16,500 feet. Going a little higher, they could see the upper slopes of the mountain. Their first view of the face we would climb was sobering: hard, shiny ice slopes with overhanging glaciers.

A few days later Irene, Vera K., and I followed a trail made by Vera W., Margi, and Annie, and chose what we hoped would be a safe site for Camp II at 18,500 feet. Of the nine climbers who'd died on Annapurna, six had been hit by avalanches at Camp II. We set up our tents on a level area, forty feet in diameter. A steep ice wall would, we hoped, shield us from avalanches. To further protect us, Lobsang, our head Sherpa, said prayers and sprinkled holy rice around our campsite.

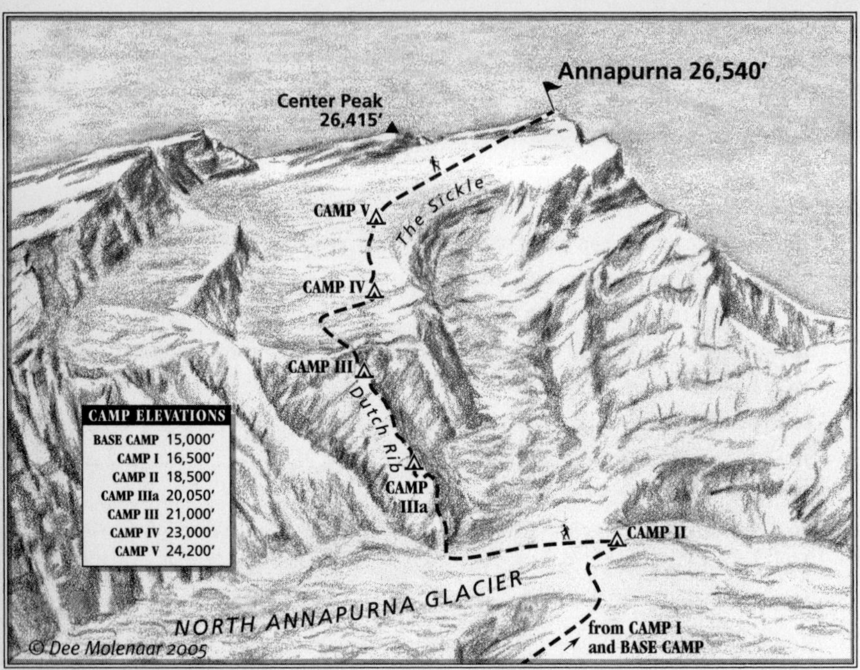

We spent ten days carrying loads to acclimatize and stock the camps. The team fell into a pleasant rhythm of hard physical labor during the day and storytelling or singing at night. Cuddled in our sleeping bags, all crushed into one tent, we had endless discussions of husbands and boyfriends, how much we missed them, and graphic details of what would transpire when we saw them again. On my past trips, when I was the only woman, the incessant risqué banter about women and sex had upon occasion been annoying to me; on our women's trip the tenor of many conversations was surprisingly similar, but lots more fun.

Then the question arose of who was going to lead on the steep slopes ahead. Up to this point, three teams of three members each had equally shared the enjoyable, challenging work of leading and the routine work of carrying loads. But I believed that for safety, the higher, more dangerous slopes should be climbed as quickly as possible. I decided that our four best ice climbers would lead and everyone else would follow with loads. I didn't ask for anyone else's opinion on this important decision, fearing that the team would choose the more democratic but possibly less safe plan of everyone having her turn to lead.

At breakfast the next morning, it came out that everyone was dissatisfied with my plan and disgruntled that I hadn't consulted them. For three hours we sat around the breakfast table sharing our concerns, our hurt, our anger, and finally, our caring and love for one another. I agreed that from now on, I'd get everyone's input before making a critical decision. After much discussion about who would lead on the ice face ahead, the team reached the same conclusion I had: it would be our four best ice climbers—Vera K., Piro, Liz, and Alison. But this time the plan was based on consensus, not on an order from me. I realized I couldn't always do what everyone wanted, but I could consider their input regarding important decisions. I was reminded that good leaders are good listeners.

Before heading into the danger zone, we held a ceremony asking the mountain spirits for a safe, successful ascent. Chanting and burning pungent juniper, the Sherpas raised brilliantly colored Tibetan Buddhist prayer flags high above Base Camp. At that moment, the sun emerged and the summit of Annapurna appeared above the clouds. A good omen! We would begin our climb of the highest, most dangerous slopes of Annapurna as a united team, with the blessings of the mountain gods.

The next day, Vera K. and Piro won the coin toss to be the first to tackle the steep ice slopes of the Dutch Rib. As Piro wrote in her diary (translated below from her native Hungarian):

> We reach the Rib. Snow is bloody deep at the bottom—have to lean into the slope to pack it down for each step—even have to lift my leg with one hand under my thigh. I'm breathing like a terminal cardiac case. I turn my head and I can see the black summit pyramid 6,000 feet above me. Wow this is really

it—I can't believe where I'm climbing. It's so great I let out a yell of sheer pleasure. Around noon we decide to go down and leave the rest of the fun for Liz and Alison tomorrow.

That night a fierce storm blasted in, and Liz and Alison had to wait some days for their fun. The snow was falling so heavily that we had to shovel off our tents every half hour to keep them from collapsing. The climbers at Camp II had little choice but to abandon camp and beat a perilous retreat to Camp I as avalanches thundered all around them.

Safely down the mountain, we had a party to celebrate Margi's twenty-first birthday, Joan's fiftieth, and all we'd accomplished so far. We were halfway up Annapurna and, even better, we were all still friends and in good health. During the party I noticed Annie helping our handsome cookboy, Yeshi, in the kitchen. She had been spending a lot of time with him and I had to wonder if they were going along with the expedition rule I'd set against member-Sherpa romance. (Their relationship turned out to be more than just a high-altitude attraction; Annie and Yeshi were married a few months after the climb.)

When the storm ended we had to dig out our buried gear. Then Vera K. and Vera W. led steep ice to establish Camp IIIa, a temporary staging area for the difficult climb along the narrow ridge crest to the permanent Camp III.

On the brilliantly clear morning of September 25, Margi, Annie, and I carried the first loads from Camp II up to Camp IIIa. The beginning of the route was easy. Strolling across the gentle glacier, the Himalaya resplendent around us, I couldn't help belting out a few verses of "The Sound of Music." Then we reached huge blocks of ice, evidence of past avalanches. A chill of fear spread throughout my body and I stopped singing. We picked up the pace and soon we were trying to run uphill in soft snow among piles of avalanche debris. I was shivering with fright when we reached the bottom of the Rib and the end of the danger zone.

"I hope we don't have to cross that death trap too many times," Margi said, looking shaken.

A climber heads up from Camp II at 18,500 feet on Annapurna I.

"Well, we're across now. Let's focus on getting to the top of the Rib," I said, trying to calm myself and get mentally ready for climbing the 65-degree hard ice slope ahead.

Attaching my jumar ascender to the yellow polypropylene fixed rope left by the others, I took a deep breath. After months of anticipation, I was about to do my first technical climbing on Annapurna. I put the front points of my crampons in the small steps Piro had chopped and then used my arms to pull myself up on the jumar attached to the rope. Again and again I repeated these movements as I made my way up the ice. After several pitches my right bicep began to burn. The angle of the slope eased a little and I could relax and enjoy the magnificent views of complex ridges, huge ice towers, and seemingly endless ranks of ice mountains around me. I squeezed very carefully through a gap between the ice face and ten-foot fanglike icicles hanging over the face and then up a final steep ramp. My spirits soared. I had made it to the crest of the Dutch Rib!

After dropping our loads at Camp IIIa, a small level place next to an ice wall, we rappelled down, elated to be in this spectacular vertical world. We left our crampons and jumar ascenders at the bottom of the Rib to save the work of carrying them to and from Camp II.

The next day, Vera K. and I carried a load of hardware up to Liz

Annie rappels through the giant icicles below Camp IIIa on the Dutch Rib.

and Alison, who were leading along the ridge crest to Camp III. When we reached the bottom of the Rib, where we'd left our gear the day before, everything was gone—buried by an avalanche. Without crampons and jumars, we couldn't climb any higher. As we dug through the debris trying to find at least some of the lost equipment, we heard a thunderous roar. I looked up and saw an avalanche cloud far above, gathering size and speed and heading straight toward us. I froze for a second, remembering the avalanche on St. Elias that had obliterated my friends. Then we tried to sprint back down the trail we'd come up. The roar grew deafening, the cloud larger. When it was almost on top of us, Vera and I crouched down and dug our axes into the slope. I focused on pushing the axe into the slope and not being swept away, holding on for my life. I felt a tremendous wind, battering snow, and then an eerie calm. The avalanche had stopped right next to us. Vera and I looked at each other caked with snow, and at the place where the cache had been, now hopelessly buried by a second enormous mass of ice. There was nothing we could do but retreat to Camp II, humbled by the immense power of the mountain and grateful that the avalanche had spared us.

Back at camp, I began a new logistics plan taking the missing gear into account. I was lost in my calculations when I heard an ominous roar. Jumping out of my tent, I saw a tidal wave of snow and ice—an avalanche cloud at least two miles wide and growing bigger by the second—sweeping down from a peak across the valley and heading toward the glacier above Camp I. I dashed to the edge of our camp and saw that the film crew and their helpers were directly in the path of the foaming mass. Within seconds the avalanche cloud swept across the glacier and Dyanna, Marie, Christy, Joan, and two porters, Kaji and Pemba, disappeared.

After excruciating minutes that felt like hours, the cloud settled. I counted six tiny spots moving like ants whose home has been stomped. In the second miracle of the day, the avalanche had stopped just above them, although the wind was intense enough to level Camp I a mile below.

Christy wrote in her diary:

> With a great crack, a giant chunk breaks off on top of the chute across the valley. At first we're delighted. Dyanna starts filming the avalanche as it rushes down, growing menacingly. Suddenly I realize we're going to get it. We're miles away, but that thing is moving like fury. My God! I look for a depression. Can't find one—dive for the glacier and thrust my axe in. Marie screams, "Christy, what do we do?" "Get down." It's the last thing I can say before the rush of wind begins. The first blast picks me up from flat on the glacier and carries me twenty feet through the air. Next comes tons of snow and ice driven by eighty-mile-an-hour winds. Miraculously my hat is still on my head, and I snatch it to cover my face and breathe through. Every bit of exposed skin is abraded by the driven ice and hurts like hell, but I just keep breathing. As the wind dies to about twenty miles an hour, I

Irene and Vera W. at Camp II as an enormous avalanche plunges past them toward the film crew above Camp I. *(Copyright 1979 by the National Geographic Society)*

> look out to pure white. My glasses are plastered inside and out. I take them off, but it's still a wind-driven whiteout.
>
> When it finally dies down, I check people. Joan and the porters Kaji and Pemba are fine. Marie is screaming now, pouring out her terror, so I know she is alive and well. Dyanna is quieter and harder to find. She was blown into a crevasse. One or two feet in either direction would have been straight down—we couldn't see the bottom. Luckily she landed on a snow bridge only six feet down. The power of this mountain overwhelms me.

During those terrifying moments when the enormous avalanche cloud covered the glacier, I realized with horror that six people might die and I had been responsible for bringing them here. I regretted ever coming to Annapurna, ever wanting to climb mountains, ever dreaming of women going to 8,000 meters. Let them live, I prayed, and we'll go home and leave Annapurna in peace.

That night I awoke at 3:00 A.M., agonizing about whether we should go or stay. I knew that of the more than ninety climbers who had attempted the mountain, only eight had reached the summit, and nine others had died trying. How had we had the hubris to tempt fate? Memories of my friends and lovers who had been sacrificed to the fury of the mountain gods swirled through my brain. I no longer had any desire to reach the summit of Annapurna. All my ambition had been swept away by the ferocity of the avalanches. Certainly my teammates would feel the same.

But the next morning, I learned they did not. When I suggested to Vera K., Piro, and Annie that we have a meeting to talk about leaving, I was met by a very long silence. Finally Annie said, "I don't feel like talking about it. I just want to climb this mountain."

I went to Irene and Margi's tent, where they were more willing to express their fears. "I've been climbing for twenty years," Irene said, "and I've never been on a mountain that is so unstable. The avalanches are completely random. Sometimes it seems almost immoral to keep going."

"If an avalanche as big as that one came down from the glacier above us, we'd be hit here at Camp II," Margi said. "This camp could be lethal."

"I agree," I told them. "I feel like we did everything right, except picking our peak. I know we've got the ability to climb this mountain, but I don't know if we can keep walking under these avalanches day after day for weeks without someone getting killed. It's just not worth it."

Then Irene voiced a stubborn conviction that prevailed in the group. "I wish we'd picked another peak, too, but we're here now. I guess we'll have to make the best of it."

No one came out and said she wanted to leave; the unspoken decision was to stay, and as the leader, I had no choice but to remain with my team.

For two more weeks we persevered through storms and avalanches and even a successful Sherpa strike for higher wages. By the first week in October we had established four camps and stocked them with enough food and gear to attempt the summit.

The question that had hovered unspoken during the entire trip now had to be answered: Who would try for the top? I had assumed that by the time we were ready for the summit push, natural selection would have made the teams obvious. But there had been little attrition during our month and a half on the mountain. Joan had been ill and Liz had to return to her job. Since the big avalanche that almost killed the film crew and their helpers, I no longer cared about reaching the top myself. My focus was doing logistics as well as I could so we could get up and down the mountain as quickly and safely as possible. Still, seven women and most of the Sherpas wanted to be on summit teams.

This was one decision that I as leader would ultimately have to make myself. As I got everyone's input, I remembered the bitterness surrounding the summit-team decision on Everest. I'd hoped for cooperation and harmony on our expedition, but I learned that all climbers, regardless of gender, are at the mercy of both nature and human nature. Difficult as it was to admit, women could be as competitive and edgy as men when stretched to their limits in the high, thin air. Although everyone agreed that if anyone reached the top it would be a victory for us all, each climber wanted to be that "anyone."

Decades later, listening to lengthy tape recordings of conversations during which we negotiated the summit teams, I hear myself patiently trying to find consensus amidst a morass of emotions. Hearing my calm voice, I realize my early training with my mother and grandparents had been excellent preparation. Indeed, the emotional content was similar to that of my childhood: intense love, anger, and frustration all crammed into a very small space.

In the dark, lonely hours of the night, I thought about the qualifications of each team member and made my plan. I selected Irene, Vera K., and Piro as the first team and Vera W., Alison, Margi, and Annie, accompanied by Mingma and Chewang, two of our Sherpa high-altitude porters, as the second. Margi volunteered to go up with two of the Sherpas to establish Camp V. The team was generally content with this plan.

On October 13, I accompanied the first summit team partway up to Camp IV. At about 22,000 feet I hugged them good-bye. I would descend by myself to Camp III. This would be my high point on Annapurna. I still hadn't seen the brown hills of Tibet, but all that mattered to me now was that someone get to the top of Annapurna and, more important, that all of us get back down alive.

As I descended I heard a new sound, a honking above me. I looked up and saw a flock of bar-headed geese flying from Tibet to India over Annapurna. As I watched them circle the summit and resume their flight south, my spirits rose. I was moved and encouraged at the sight of other living creatures in this high, alien world.

The next day, Margi hobbled down from helping establish Camp V. Her toes and heels were frostbitten—she would not attempt the summit. She told me that Mingma and Chewang had insisted on trying for the top with the first summit team. The team's radios were broken, so I couldn't reach them to discuss the matter.

Irene reaches Camp III, at 21,000 feet, on her way to the summit of Annapurna I.

Summit morning, October 15, 1978, the climbers got up at 4:00 A.M. While putting on her crampons, Piro noticed a hole in her glove and that her right index finger was frozen solid. Dreading the possibility of mutilation or amputation, Piro put her finger in her armpit and leapt back into the tent, crampons and all, to thaw her finger, her summit bid abandoned. Irene, Vera K., Mingma, and Chewang headed for the top of Annapurna I at 7:00 A.M.

Meanwhile, I was at Camp III with the second summit team. After much deliberation, Annie decided to go down rather than try for the top. She was concerned that without Margi or much Sherpa support their team wasn't strong enough, and she knew she had many more years to go high. I shared Annie's concerns, but Alison, who had climbed higher than 26,000 feet without Sherpas or oxygen, was confident.

"We can do it," Alison said. "Come with us, Annie."

"I bet the first team will make it," Annie said. "A second ascent isn't worth the risk. In a year, it won't matter who got to the top."

Alison wasn't deterred. "This is what our expedition is all about—a real all-women's summit team," she said. "Please come."

But Annie held firm to her decision.

In the late morning Marie, who was watching the summit team from Camp I with binoculars, came in over the radio. "They're doing it," she said. "Four hours more to the top at this rate. It's so beautiful to watch. Deep blue sky, a plume of snow blowing off the top, and our friends way up there. A great day. Over and out."

I heard Marie, but was more focused on trying to dissuade Vera W. and Alison from going up. "If anyone reaches the top, it will be a victory for us all," I said.

But Alison wanted her own chance. "We've risked our lives carrying loads across those avalanche chutes for the others," she said, cinching her pack closed. "We're adults and we've earned the right to try for the top ourselves."

"Please be careful, Alison," I said. "And promise to turn back if it gets dangerous." I hugged Vera W. and Alison good-bye. They shouldered their packs and set off for Camp IV.

As they had taken the Camp III radio up with them, I decided to head down right away so I could stay in touch via the Camp II radio. Margi and Annie would follow me down later.

I set off by myself, my backpack stuffed full, nervous about going down the six-inch-wide crest of the Dutch Rib alone. I walked down backward along the very narrow ridge, keeping the fixed line taut and trying not to look at the sheer drop-off on either side. Glancing back over my shoulder, I saw a sinuous white tightrope stretching on and on, seemingly forever. I couldn't help thinking, *What's a nice Jewish girl from the Midwest doing at 21,000 feet, going down a knife-edged ridge all alone?*

Descending the last ice pitch above Camp IIIa, I fell waist-deep in a crevasse full of powder snow. Afraid of falling even deeper, I clipped my jumar to the rope and thrashed to the surface. I felt like crying, but continued down the steep slopes on the side of the Rib. Finally, blessedly, I spotted someone in the distance. It was Christy, who had very kindly come back up to meet me.

I summoned what strength I had left to yell down to her, "Did they make it?"

From Irene's diary of the summit day:

> We don't talk as we climb higher. All of our energy and concentration go into the steady, monotonous plod that is taking us toward our goal. There is still no wind, but we can see plumes of snow blowing off the summit in the winter gale. I think of my family and friends. Their love is a steadying force, easing my way up the mountain.
>
> Just below the summit pyramid the snow is again very deep, and our pace drags. The bands of rock below the summit that I had worried about for months turn out to be no problem. We walk right over them, our crampons grating on the sandstone and we gain the crest of the windy, corniced summit ridge. But where is the summit? Chewang gets summit fever and starts racing along the ridge trying to determine the highest point. We traverse three or four bumps, and finally we are there.

From Camps I and II, the rest of the team saw through binoculars the dramatic sight of four tiny figures reaching the top of Annapurna as a plume of blowing snow streamed from the summit. A huge cheer went out on the radio between camps.

At 3:30 P.M. on October 15, 1978, our team reached 26,540 feet, the top of the planet's tenth-highest mountain. The four successful climbers—Irene, Vera K., Chewang, and Mingma—unfurled an American flag, a Nepalese flag, and A WOMAN'S PLACE IS ON TOP flag, all held together by a SAVE THE WHALES pin. Our ascent was the first time any American or any woman had climbed Annapurna I, and set a world altitude record for American women.

When Christy nodded her head yes in response to my shouted question about whether we'd succeeded, I sat down in the track and cried. Physically and emotionally exhausted, I felt a mixture of triumph for the summit, relief at having made it down the Dutch Rib and across

Irene on her way to Annapurna I, with Dhaulagiri in background.

the avalanche slope for the last time, and, most of all, joy in knowing that after all the years of dreaming, planning, and preparing, we had climbed Annapurna. A woman's place was indeed on top of Annapurna or anywhere else she chose to go!

Christy came up, gave me a hug, and took my pack. I was very grateful to her and walked slowly down behind her.

Chewang and Mingma hurried back to Camp II the next day and announced, "Annapurna summit reached. Let's go to Kathmandu and have a party!" It was a good idea, but first the second team would try for the top.

On October 17, two days after the first team's success, Vera W. and Alison set off from Camp IV. The weather was clear, and they were confident and in good spirits, according to Irene, Piro, and Vera K., who saw them off and stayed up at Camp III in support. Marie and

Christy gives Annie the joyful news that Irene, Vera K., Chewang, and Mingma have reached the summit. October 15, 1978. *(Copyright 1979 by the National Geographic Society)*

Dyanna, who were filming them from Camp I, reported that they were moving well. By dinnertime Vera W. and Alison were at the base of a small ice face just below Camp V and the face was too heavily shadowed for filming. The film crew went into the mess tent for dinner, estimating that the two climbers were about twenty minutes below Camp V.

Vera and Alison didn't contact us on the radio that evening, nor did they respond to our calls. I slept fitfully that night, awoke early, and continued calling them. We scoured the slope around Camp V with binoculars and called over and over. Nothing. Instead of feeling happy that we'd climbed Annapurna I, we were consumed with worry. Why weren't they answering? The possibilities were so awful that we didn't dare voice our fears.

Finally Mingma and Lakpa Sherpa agreed to go back up to see if Vera and Alison needed help. Despite their fatigue, they would go from Camp II to Camp IV in one day and give us a report from Camp V the following morning. A few hours after they headed up, their frantic voices came in on the radio.

Yeshi translated. "Mingma and Lakpa see a red jacket by Camp IV. They think it's Alison." Picking up my binoculars, I looked to the left of Camp IV and could just make out a red blur. Yeshi continued to translate and we heard what we had dreaded most: Mingma and Lakpa had found Vera's and Alison's bodies.

I felt a physical jolt of pain course through me, but I needed to stay calm to do my job. I reminded myself to breathe, and in a somber voice announced the dreadful news on the radio. Only when I had comforted the others did I go back into my tent and cry for Vera and Alison, disconnected images of them running through my mind. I thought of Vera in her sunny kitchen, Vera dancing up a rock face with grace. And Alison. I saw her in Afghanistan, so happy, arm in arm with her husband, Janusz, playing with the children on the way in. I remembered when she fell just below the summit of Noshaq and stopped herself just above a drop-off in the dark. But this time she had been unable to save herself.

Alison and Vera had climbed hundreds of mountains. They both knew that a fall is an always-present danger. One foot placed insecurely just an inch from safety, a shift of weight, a slip; a rockfall, an avalanche, a brief loss of concentration could, in an instant, take a climber from life to death.

As in the Pamirs and on Trisul, I tortured myself with all the ways it could have turned out differently. I wanted to change the climbers, the mountain, our daring to try to climb high. And in the midst of my sorrow came anger toward them for leaving us, and then guilt for feeling angry.

The Sherpas came down and reported that Alison's body was still tied to the rope, which led into a crevasse where they believed Vera had fallen. They were stopped from reaching the bodies by another big crevasse.

We judged it too dangerous to go back up to bury their bodies, so Vera and Alison stayed on Annapurna. We packed up our huge, heavy loads, and with even heavier hearts we cleared the mountain and made our way down to Base Camp.

Once there, we added Vera's and Alison's names to a memorial stone on which were engraved the names of the seven other climbers who had died on this side of Annapurna. Scarcely able to see through my tears, I chipped out the letters on the stone with a hammer and screwdriver. Their names would stay forever in the rock, looking toward the summit they had so hoped to reach.

OCTOBER 17, 1978
VERA WATSON
ALISON CHADWICK-ONYSZKIEWICZ

Our last evening at Base Camp, just at dusk, when the mountain was shrouded in fog, we had a memorial service. We shared our memories of Vera and Alison, their warmth and laughter, their energy and enthusiasm, their integrity and idealism. We all locked arms and sang the old Shaker song, "'Tis a gift to be simple, 'tis a gift to be free, 'tis a gift to come down where we ought to be . . ." And the Sherpas began to chant, *"Om mani padme hum, om mani padme hum . . ."* "The jewel is in the lotus, the spirit is in the universe, the spirits of Vera and Alison are with Annapurna."

I chip Vera's and Alison's names into the memorial stone at Annapurna Base Camp.

Chapter 20 ▲

Meeting My Father

Chicago 1960

I'm running and my father is next to me: the Gestapo and their dogs are chasing us. The sound of barking and of boots hitting pavement gets louder. Looking up at my father, I see that his face has no features. Shaking my shoulder gently, Grandpa wakens me from the recurrent nightmare I've been having ever since last month when I wrote my term paper on the Holocaust. It's time to get up for school.

Walking home that day, I rehearse a conversation with my grandfather. I will persuade him to tell me about my father. After I finish my homework, I go out and sit next to Grandpa on the porch. The blood pounds in my ears as I blurt out, "Please tell me. What's the matter with my father? Does it have something to do with the Holocaust? What happened between him and my mom?"

There's a long silence.

"You shouldn't worry about such things. They have nothing to do with you," Grandpa says.

"You've got to tell me. I'm having nightmares. I need to know."

The red center of Grandpa's cigarette gets brighter and darker as he slowly sucks on it.

"You're fifteen now. I guess you're old enough," he finally says. "Your father lives in New York. He sends money for you each month."

I'm thrilled, shocked, and terrified all at once. "I want to meet him," I say. My dad knows about me and sends money. He must love me.

"You don't need him," Grandpa says.

"He's my father and I do so need him."

"You can meet him when you're older," he says, and pushes himself up off his chair.

A few weeks later, to my total astonishment, Grandpa tells me that he's gotten in touch with my father through the lawyer who sends the child-support checks. My father will meet me in Chicago next month. I'm in heaven. My dream is coming true. I have a father like everyone else. When the day finally comes, I change my outfit three times. In our 1954 blue Plymouth sedan, Grandpa drives Mom and me to the Hyde Park home of my father's friends Martin and Mitzi Marx. As we drive, I feel like I'm hovering on the ceiling of the car, watching myself, the heroine in a movie, going off to meet her loving, long-lost father, with Beethoven's "Ode to Joy" playing in the background.

Mitzi, a short, buxom woman, answers my timid knock and ushers me in. My mother leaves with my grandfather, saying they'll be back in two hours. My legs trembling, I walk slowly into the living room, where a slender, smiling man wearing glasses is waiting on a couch encased in clear plastic.

He stands up and we shake hands clumsily. Mitzi brings in a plate of linzer torte and she and my father chat about people I don't know. Saying we should have time on our own, Mitzi leaves the room. I nibble at my pastry, afraid to speak or look at my father. But I have to say something. Should I ask about the weather in New York? Or where he's been all these years? Finally I ask him what he does. In a thick German accent, he tells me he's an anesthesiologist and then describes a new technique he's developed. I don't really understand what he's talking about, nor do I completely believe that this foreign man with the guttural voice is my long-dreamed-of father. He tells me that he drives ninety minutes each way to his job at a veterans hospital. He keeps talking, about his art gallery in Manhattan, his involvement in the Democratic Party, the folk-dancing club he started. As he describes his life, it sounds amazingly full and rich compared with mine.

Meeting my father at age fifteen.

Then he hands me a photo of an attractive blonde woman holding two cute chubby toddlers. "This is my wife, Mary, your sister, Claire, and your brother, Jonathan, on the front steps of our town house in Queens," my father says. I look carefully at the woman and two children, children who actually live with my father every day. Who is this woman? My stepmother? Who are these children? My brother? My sister? *It's more than I can take in. What does he find with this family that he didn't find with my mom and me? Does he love these two unfamiliar children more than he loves me? I shove these thoughts aside and focus on the idea that this could be my new family. We talk a little more about his life and the possibility of my coming to New York to visit him.*

I hear a knock on the door and hand the photo back. My father gives me an awkward kiss on my cheek. Our two hours are over for today; tomorrow I will see him one more time before he returns to New York.

First Up Bhrigupanth ▲

1980

IN THE MONTHS AFTER our return from Nepal, Vera's and Alison's deaths overshadowed our climbing Annapurna in my mind. The media's reaction had been similar: our reaching the summit was a small note buried in the local newspaper; the tragedy made front-page headlines around the world.

In the calm light of day, I felt that in spite of our terrible loss we had achieved many of our objectives. We were the first American women ever to climb an 8,000-meter peak and we had reached that altitude on a mountain more dangerous and difficult than we could ever have anticipated.* Team members told me I had done an outstanding job of leading and making decisions during the long and complex endeavor. Irene and I, along with her husband, Colin Miller, and my boyfriend, John Percival, were honored to be invited to the

*According to a 2005 update on climbing 8,000-meter peaks in *Appalachia* magazine: "Climbing Annapurna has proved to be neither easy nor safe. It is the least visited of the 8,000ers and has the dubious distinction of the lowest ratio of successful ascents to deaths of any of them."

Irene Beardsley Miller; Sarah Weddington, President Carter's advisor on women's issues; and me in the White House after the Annapurna climb, 1978.*

White House, where Sarah Weddington, President Carter's advisor on women's issues, congratulated us for the Annapurna climb.

However, at three in the morning, I'd awake with questions whirling like demons through my brain. Should we have left after the big avalanche? Would another summit plan have made the difference? Could I have stopped Vera and Alison from trying for the top? Why had we ever attempted such a dangerous mountain? And then my slow progress toward equanimity was shattered when *National Geographic* sent me a copy of a letter they'd received warning them against publishing my article about the Annapurna expedition. Signed by a woman I didn't know, the letter had a return address in Berkeley, California. She wrote:

> To claim this as a woman's achievement is bald-faced racism of the type outlawed in America since the Civil War. Had the men on Annapurna been Americans instead of Nepalese, no one would have gotten away with a claim that this was an achievement by and for American women. This was a coeducational, cohabiting expedition in which women slept with men who were plied into sexual relations with booze and the escapade ended with one woman marrying a Sherpa. . . . *National Geographic* must recognize this misrepresentation, or be prepared for criticism far exceeding anything received on your carefully researched stories to date. Your magazine has a time value, a ring of the truth. Please don't send your credibility down the river.

---

*These historic T-shirts are still available by writing Himalayan Trekking Class, P.O. Box 5455, Berkeley, CA 94705, or ordering online at www.arleneblum.com/t_shirts.html

Gil Grosvenor, the editor in chief of *National Geographic,* responded firmly, refuting in detail the accusations in the letter. "That they had Sherpa assistance is not unusual, and does not compromise the effort," he wrote. "Your remarks relative to the conduct of the expedition members raise questions of personal propriety, based on hearsay, that seem to me to be beneath the questions at issue. . . . I am confident Miss Blum's article is sound." The editor cautioned me that one or more Berkeley climbers were thought to be behind the letter and they might make trouble for us in the sensationalist media.

The letter's aggressive tone made me feel as though I had been physically assaulted. Who was the woman who wrote the letter and what did she have against us? Who else was involved? In eloquent imaginary rebuttals, I pointed out that no one denigrated the British first ascent of Everest because the Sherpa Tenzing Norgay went to the top with Edmund Hillary, or our American Bicentennial Everest Expedition for employing forty Sherpas. Regardless, the letter was an ugly reminder that women's climbing was still threatening to some people, and I was deeply troubled to think there were those who wished us ill.

At the time I was happy to be living with John Percival, my centered, tranquil boyfriend. I cared deeply for him, but my nighttime anxiety and daytime obsession with writing my Annapurna book took their toll on our relationship. In fact, during this same time, the marriages and relationships of many other Annapurna team members ended. It is not an unusual phenomenon. When one person in a couple has such a life-changing experience, it can be destructive to the partnership. After the 1963 first American ascent of Everest, for example, the marriages of many of the climbers ended. Only when John left to take a job at Lake Tahoe did I realize how much I'd lost, but our separation turned out to be irrevocable. I wrote him long, pining letters, but there was no going back. I had to get on with my life.

Still grieving for Vera and Alison and regretting the end of my relationship with John, I began to think about leading a women's expedition to a smaller Himalayan peak. This time, I vowed we would choose a safe mountain and, utilizing our past experience, we would do things right. Just women; no high-altitude porters, no Sherpas. We would silence the doubters who said women couldn't climb high on their own.

I was pleased when Piro and Christy from Annapurna decided to join me, Christy making her eagerly anticipated transition from Base Camp manager to climbing-team member. Next, we had to choose a mountain. A friend told us about Bhrigupanth, an unclimbed, 22,300-foot ice peak in an area that was just opening to Western climbers in the Garhwal region of northern India. At the bottom of the nearby Gangotri glacier, a torrent of water foams out of a giant ice cave called Gomukh, "the mouth of the cow." Gomukh is the source of the Bhagirati River, a major tributary of the holy River Ganges and a sacred destination for pilgrims from all over India. After applying for a permit to climb Bhrigupanth in the spring of 1980, we looked for several strong young climbers for our team, wanting to share this opportunity with the next generation.

Putting the word out in the climbing community, we recruited Nancey Goforth, twenty-seven, Susan Coons, twenty-eight, and Penny Brothers, twenty-three, all skilled mountaineers for whom this would be a first Himalayan expedition. Nancey, a biologist and veteran Outward Bound instructor, had recently graduated from Evergreen State College in Olympia, Washington. Susan was a poet and sleep researcher at Stanford University, and Penny, a chemistry graduate student at Stanford, was from Auckland, New Zealand. The friend who'd told us about Bhrigupanth also put us in touch with Rekha Sharma, a twenty-one-year-old textile designer from New Delhi who had climbed in the Himalaya since age thirteen. At Rekha's suggestion, we invited a climbing partner of hers, Rajkumari Chand, a twenty-nine-year-old school inspector who had grown up in the Indian hill town of Almora. Now we were eight.

The nonclimbing, ninth member of our team was Barbara Drinkwater, a sports physiologist. Barbara had studied the Annapurna team before that climb, but on Bhrigupanth she would do the first-ever field studies of women's adaptation to high altitude. Along with Piro, who would again be our team physician, Barbara would monitor how our bodies adapted to cold and changing elevation as we made our way up the mountain. She could then compare her data with the many existing field studies of male climbers.

Meanwhile, I worked on my Annapurna book. Midway through the project my publisher, Thomas Crowell Company, was bought by Lippincott. My new editor told me that in order to produce a dramatic bestseller, they were assigning my book to a ghostwriter. I was dismayed, imagining the potential for inaccuracy and sensationalism.

Fortunately, just then I met Luree Miller, a warm, sympathetic scholar and world traveler, and the author of *On Top of the World: Five Women Explorers in Tibet.* Luree adamantly supported my taking the Annapurna book back from Lippincott. With her encouragement, I finished *Annapurna: A Woman's Place* and then resold it to Sierra Club Books, which was happy to publish the story in my own words.

As I continued my research with Bruce Ames at UC Berkeley, my focus shifted from Tris and Fyrol to other halogenated hydrocarbons, all of which are long-lived, accumulate in body fat, and are harmful to people and the environment. I was especially concerned about three fumigants used to protect foods from insects and spoiling: methyl bromide,* ethyl bromide, and dibromochloropropane (DBCP).

The Oil, Chemical, and Atomic Workers Union (OCAWU) asked my advice about male chemical workers at Occidental Petroleum, a disproportionate number of whom were sterile. When I learned they worked with DBCP, I suspected this fumigant could be the reason. It causes mutations and these changes could accumulate in the fatty cells that produce sperm. To learn more about the effects of this chemical and to try to help the sterile DBCP workers, I took a second job as a consultant for the OCAWU. After testifying about the hazards of the chemical at EPA hearings, I was delighted to find my testimony the

*In 2005, the United States sought an exemption from guidelines set in the 1987 Montreal Protocol for the phase-out of methyl bromide, a potent ozone-depleting chemical used on food crops. Originally developed during World War II as a nerve gas, it also causes reproductive problems, and exposure to it has been linked to an increased risk of prostate cancer among pesticide applicators and farmworkers.

basis for a front-page story in the *Wall Street Journal* questioning the fumigant's safety. In 1979, DBCP was banned from most uses.

While I was working on DBCP, a group of students at UC Berkeley asked me to help them put together a course on cancer. Together we designed a popular class called Cancer: Current Theories on Cause and Prevention. We invited experts to lecture on environmental causes of cancer such as smoking, chemicals, and radiation; the disease's diagnosis and treatment; and how to change our lifestyles and society to reduce its incidence. Nearly a thousand students and community members attended the class; many had a personal connection to someone with cancer. Sadly, the subject was becoming one of great personal interest to me, too.

Joan Firey, whose low energy had persisted after the Annapurna expedition, learned she had a rare type of blood cancer shortly after our return from Nepal. When I went to Seattle to visit her, Joan showed me a series of striking paintings she'd made of mountains in the North Cascades.

"These are gorgeous," I said. "How can you do all this now?"

"I'm tired, but my spirits have never been better," she said. "The cancer is teaching me a lot. I'm focusing on what's really important—my family and my art. I'm spending time with my kids and painting more than I have in years."

"You're amazing, Joan," I said quietly as I hugged her good-bye.

"I'm going to beat this," she said, hugging me back. "And even if I don't, it's one of the best learning experiences of my life."

Shortly after I got home I got a call from Margaret Young, my Denali teammate. She asked me to visit and tell her about this cancer business. I assumed she wanted to know about Joan, but when we got together Margaret told me that she herself had liver cancer. She knew I taught a cancer course and she wanted more information than her doctor had given her. I had to tell her the sad truth, that people with liver cancer usually don't have long to live.

"I guess I'll have to make every day count," Margaret said in her matter-of-fact way.

Margaret and Joan each lived about one more year. Spending as much time as I could with each of them, I was inspired by their courage and optimism, their positive energy and resolve to lead the last year of their lives in the best way they could. When they died, I vowed to appreciate the finiteness and preciousness of my own life.

At that moment, leading an international women's team in attempting the first ascent of Bhrigupanth was what I most wanted to do. The mountain was about four thousand feet lower than Annapurna, which meant we didn't need oxygen or high-altitude porters, and preparations were relatively simple. Rather than ship our food and gear ahead, we packed everything in duffels and checked them on our flight to India. Fund-raising was also a nonissue. Our three younger members were awarded grants from the fund for women climbers that we had established in Vera's and Alison's names. The rest of us paid our own shares—about $600 per person plus airfare.

When we awoke at the YWCA on our first morning in Delhi, May 13, 1980, it was 108 degrees in the shade. Penny wrote in her diary:

> A ceiling fan labored to move the oppressive air as we lounged in various states of nakedness on the vintage furniture in the room, our bodies glistening with sweat. It could have been a scene from a thriller set in an exotic locale, but our agenda was suggested by dozens of duffel bags bulging with ropes, climbing hardware, and freeze-dried food.

Faye Kerr, my good friend from Denali, and I were planning to meet following the Bhrigupanth expedition for a trek in Ladakh, a Buddhist region in northwestern India. As we loaded our bus to leave for the mountains, I received a phone call from the friend who was traveling in southern India with Faye. In a heavy voice, he told me that Faye had gone to the hospital with acute dysentery the day before, and had died a few hours later.

I was overcome with an all-too-familiar sorrow and disbelief; Faye so strong and safe in the mountains, suddenly dead in a hospital in Madras. As our bus twisted its way up into the Himalaya, all I could think of was kind, gentle Faye; I remembered her delight in the mountains and how, regardless of physical discomfort, wherever she found herself was the perfect place to be. I didn't have the heart to go climbing, but I was the leader of eight other women who did, and we were on our way with all our gear and an unclimbed peak awaiting us.

A long day's ride took us from the plains to the foothills on hairpin turns, then to Uttarkashi, where we joined the Ganges and then the Bhagirati. I mourned the loss of Faye as we journeyed, along with Hindu pilgrims from all over India, up the river north toward its sacred source at Gomukh. Reminded yet one more time of the nearness of death, I resolved to appreciate every minute of my life.

Arriving in Gangotri on May 20, we joined hundreds of pilgrims who were chanting prayers, ringing bells, and filling jars with holy water from the Bhagirati River. Rekha arranged for a *puja* ceremony to ensure us a safe, successful climb. The priest, dressed in a white loincloth, handed us incense and herbs, painted scarlet *tika* marks on our foreheads, sprinkled rice on our heads, and tied saffron-colored holy threads around our wrists. As the sun sent shafts of light deep into the gorge leading to our mountain, I said a special prayer for Faye. Finally the priest gave us a packet of holy rice to carry to the summit of Bhrigupanth and then, upon our return, toss into the river. We were now in the care of the mountain gods.

After a three-day hike, we reached our Base Camp in a lush meadow of wildflowers at 14,000 feet. Fifteen porters (compared with the two hundred on Annapurna) had helped transport our gear. On the mountain itself, we would carry our own loads. Our first good look at Bhrigupanth showed a seemingly vertical face lined with avalanche gullies and no obvious route. We'd have to go much higher to see if there were a possible route on the back side of the mountain.

After we cast lots to determine who'd get to lead first, Piro, Rajkumari, Rekha, and I won. At dawn we set off, crossing a loose moraine, a huge frozen lake, more moraine, and snow. Trudging upward at 16,000 feet, I was delighted that I wasn't breathing hard.

Mist replaced brilliant sunshine as we ascended. At last, a view

Garhwal, India, 1980. Bhrigupanth team at Base Camp, with the ethereal mountain in the background. *(From left, standing)* Nancey Goforth, Rekha Sharma, Barbara Drinkwater, Penny Brothers, and Christy Tews. *(Front)* Rajkumari Chand, Susan Coons, Piro Kramar, and me.

of the hidden flank of the mountain unfolded. A broken icefall and then a steep couloir led to a saddle between Bhrigupanth and the enormous monolith of Thalay Sagar.

Two days later, Christy, Susan, Nancey, and Penny reached 18,000 feet and established Glacier Dome Camp on a flat bulge of ice below the wild face of Thalay Sagar. At that altitude, there is less than half as much oxygen as at sea level and little protection from the tropical sun. The snow melted into knee-deep slush by 10:00 A.M., so we adopted a pattern of rising at 3:00 A.M. and leaving in the dark by the light of our headlamps to avoid the midday heat.

Carrying our own loads was emotionally and logistically simpler than managing Sherpas and porters as we'd done on Annapurna and Everest. I felt a quiet satisfaction that we were making all the decisions and doing the hard work ourselves. In the afternoon, back at Base Camp, we melted snow, cooked, and took part in physiology tests for Barbara and Piro. As we acclimatized, our performance at 14,000 feet was coming close to what it had been at sea level prior to the trip.

Barbara was surprised that my lung capacity was actually that of an average woman and relatively small for someone of my height. I wondered if the cigarette smoke and the pollution from the nearby steel mills that I had been exposed to during my childhood could have been a factor. If there was soot on my pillow every morning when I was young, what had been collecting in my lungs?

After a week of load carrying, we were ready to ascend a 2,000-foot snow-and-ice couloir leading to a 20,000-foot saddle between Bhrigupanth and Thalay Sagar. We hoped that from there we could find a way to the top. When I asked Rekha and Rajkumari if they would like to lead on the slopes ahead, their answer was: "We'll try." From past encounters in Asia, I guessed that they were politely telling me that they hadn't yet learned to climb such slopes, which indeed turned out to be the case. So we spent a day teaching them ice-climbing techniques. Although Rekha and Rajkumari learned easily, they lacked the experience and confidence to lead and decided not to try for the top.

On June 8, most of the team headed up the couloir, which became steeper the higher we climbed. The top 500 feet were the worst—three feet of soft snow over hard ice. We fought our way upward, taking at least three breaths for each step. Just below the top of the couloir, it was my turn to break the trail. Wallowing through

Penny and Susan heading up to the col between Bhrigupanth and Thalay Sagar.

the heavy snow, I wondered what we would find in this place where no one had ever stood before. Pulling myself over the lip above the couloir, I hoped to see two yards of sloping snow on which to pitch a tent. Instead I saw below me an expanse of totally flat snow the size of two football fields. This plateau was even more implausible for being surrounded by the near-vertical faces of Meru and Thalay Sagar. As I belayed the others up, I couldn't stop marveling that we were the first people ever to set foot in this wondrous snow basin.

Piro then led down in a fast sitting glissade, during which she flew across an open crevasse before stopping at the place we would

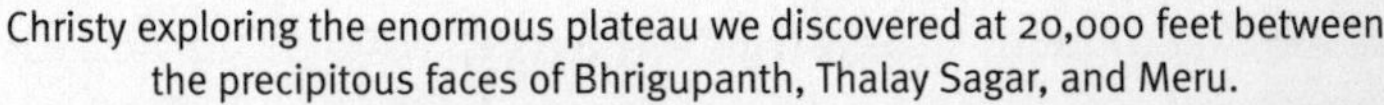

Christy exploring the enormous plateau we discovered at 20,000 feet between the precipitous faces of Bhrigupanth, Thalay Sagar, and Meru.

pitch our tents. We had discovered a horizontal world at 20,000 feet, and across from us we saw a possible route up the back side of Bhrigupanth.

Energized by our discovery, we worked hard for three days moving loads up the steep couloir to the snow basin. Then Penny, Piro, and Susan crossed the basin to reconnoiter the way to the top. They were elated when they found an intricate route through a formidable wall of ice, but the summit was still far above. Piro decided her priority was to go back to Base Camp to get ready to carry out physiological measurements on the returning climbers. I was suffering from bronchitis and was content to encourage the summit team from High Camp. Nancey, Susan, and Penny, who were then the strongest climbers on our team, would try for the top.

On June 17, 1980, the three set out at 4:00 A.M. in the bitter cold. By dawn they were above the icefall. Susan wrote about the climb:

> It was an effort, clearing the snow, kicking crampons into the ice, and swinging the axe, the cold air burning my throat and lungs. It took fifteen minutes to place a screw in the dense and brittle surface, three false starts, till I made a placement without shattering the ice. Then Nancey followed me up and took the next lead.

The wind-packed crust slowed their progress, and by noon they'd covered less than half of the 3,000 feet from High Camp to the top. They decided to bivouac in a snow scoop next to a huge rock buttress at 21,300 feet. Penny wrote that one of her best memories from the climb was that first evening in the bivy, the three climbers crammed into a four-by-six bivouac sack and two zipped-together sleeping bags:

> The red rock on Thalay Sagar glowing fierce and bright in the evening sun, the shadow of Bhrigupanth cast on the face of Meru, the thread of warmth and encouragement that connected us to Christy and Arlene just below.

The next morning, Christy and I watched from High Camp as three small figures moved slowly and steadily upward. By late afternoon only 250 feet remained, but it was over steep, rotten rock plastered together with snow. At dusk we saw the three climbers heading back down. Was this the end? Turning back only 250 feet from the summit was demoralizing. But retreat did seem the prudent course.

Penny called on the radio in the evening to say they had no food or fuel in their cramped bivy. All that night I worried about whether they should go up or down in the morning, but the decision was theirs to make.

At dawn Christy and I watched three specks emerge from the bivouac, hesitate, and head back up. They must have been very hungry, thirsty, and tired, but on the other hand they were fit and well acclimatized, and the weather was good. I admired their resolve and prayed for their safety.

Taking advantage of the frozen trail from the first day, they reached their previous high point in three hours. Then the afternoon clouds came in, obscuring our view. Christy and I kept our eyes glued to the mountain but we could see only clouds swirling past the summit. At 6:00 P.M. the clouds cleared and still we saw nothing—no one on top, no one descending. Where were they? Why were they not responding to our radio calls? They could not have reached the top and gone all the way back to the bivouac already. All I could think of was Vera and Alison. Why had I, once again, brought a team to the Himalaya? What a fool I was!

After a restless night filled with dreams of avalanches and falling ice, Christy and I awoke to a gray dawn. At first light we focused our binoculars on the bivouac camp, looking for some indication that our three friends had made it down the previous night in the dark. There was no movement. The minutes passed. The camp remained motionless. Where were they? I once again felt the torture of not knowing my friends' fate and not being able to help them. I never wanted to feel this agony again. Much as I loved climbing in the high Himalaya, I resolved never to lead another expedition to a dangerous peak. No mountain could be worth the possible loss.

Through my dread I saw a slight movement. Was it the wind? It was definitely a person. "Sooey! Sooey! Sooey!" Christy let loose her famous Iowa hog call. We screamed and hugged each other and then looked for the other two. Very slowly a second figure and then a third emerged from the bivouac tent. Nancey, Penny, and Susan were alive! And they were coming down the mountain.

A few hours later the three exhausted climbers stumbled into High Camp, where we welcomed them with gallons of hot drinks. At first their throats were so dry they couldn't speak. Eventually, in broken phrases, they shared the misery of the nights at the bivouac and the hard climbing in the deep snow. The last pitches to the top were the worst.

Penny told us of her hair-raising final lead on rock so rotten she could pull pieces off the cliff. It sounded like browsing in a library, examining first one book and then another—she would pull out a hunk, look at it, stick it back, pray, and then try to climb on it. After Penny finished her pitch and found a belay stance, Susan took the last lead, muscling her way up a stretch of overhanging rock into a patch of thigh-deep snow. Thrashing through the deep snow, she was ecstatic to make it to the summit ridge. But the ridge itself was a huge overhanging cornice with nothing underneath. After barely stopping herself from plunging through the cornice, she straddled the narrow summit and brought Penny and Nancey up beside her.

The three happy climbers hugged each other and enjoyed a misty view of drifting clouds with occasional glimpses of Thalay Sagar, Meru, and other nearby peaks. The first people ever to set foot on the top of Bhrigupanth, Susan, Nancey, and Penny left U.S. and Indian flags and a note with a kiwi drawn on it that read, "1980 Indo-American Women's Expedition to Bhrigupanth," and listed the names of our nine team members.

After a quick descent and a trek back to Gangotri, we threw the packet of holy rice we'd carried to and from the summit of Bhrigupanth into the river. Our trip had indeed been blessed. On a modest budget, without high-altitude porters, our team had safely made the first ascent of a difficult 22,300-foot Himalayan peak. Barbara completed the first field studies ever conducted on female physiology at high altitude. She found that women have somewhat "thinner" blood—and are therefore possibly less vulnerable to frostbite and phlebitis at high altitude. I felt at peace as I watched the holy rice we'd carried to the summit float out into the swirling Bhagirati waters and begin its journey to the Ganges and the plains of India.

I was not yet ready to return to my chemistry lab. Susan, Nancey, and I joined my good friends Anna Ferro-Luzzi and John and Claude Elk for a trek across the high Himalayan region of Ladakh. Geographically and culturally part of Tibet, Ladakh is located in the Indian state of Jammu and Kashmir. The peaceful Buddhist inhabitants live in a dry moonscape behind the Himalaya, where the monsoon rains seldom reach.

Hiring horses to carry our gear, we spent three idyllic weeks walking across this stunning arid land. This was my first experience of trekking in the Himalaya, getting strong and fit, and enjoying the people and mountains without risking my life.

The next-to-last day of the trek, just before sunset, I walked behind the others along the crest of a ridge. The air was so clear I could see every detail of the contorted rock faces around me, their striated layers folded and bunched like pieces of crepe paper by the continuing uplift of the Himalaya. The late afternoon sun colored the convoluted layers of rock and high snowy peaks in the distance with an ethereal golden hue that melted to pink, red, and orange as the sun set. Awed by the beauty, I sat down and enjoyed my glorious surroundings, full of peace and gratitude.

This was where I wanted to be: on a high ridge in the Himalaya. I thought of the stuffy chemistry laboratories where I worked and realized that I did not want to return there. With a smile, I said to myself, "How are they going to keep you down in the lab once you've seen the Himalaya?"

I thought about Margaret Young from Denali and Joan Firey from Annapurna, both of whom had such appreciation of life as they faced death. What would I do if I had one year to live? I wanted to see as much of these magnificent Himalayan realms as I could. I visualized walking the entire length of the Great Himalaya Range. Starting in eastern Bhutan, I could cross Bhutan, Sikkim, Nepal, and northern India and finish back here in Ladakh. Walking the Himalaya was what I would do if I had a lethal illness, and I might as well get started as soon as possible.

Full of joy at my new plan, I walked back into our camp in the courtyard of an ancient monastery. The others were dancing in the full moonlight. Two Ladakhi women in traditional dress led the dances, moving gracefully to the rhythm of a large tin can played by a young boy. I wanted to be part of the line of dancers, but I felt tall and awkward compared with the others. I sat alone in the moonlight, overwhelmed by emotions I didn't understand, tears streaming down my face. For a change, these were tears of joy. Everyone was alive and well after our climb. We were enjoying our fantastic trek. I had a plan for a new adventure. Life seemed full of promise.

I walked back to the end of the ridge where I had watched the sun set, the crest now illuminated by white moonlight. I sat down on a rock and cried once more for Faye Kerr. And then I cried for John Hall, Susan Deery, Stanley and Lucille Borgen Adamson; for Eva Isenschmid and the eight Russian women; for Bruce Carson, Vera Watson, and Alison Chadwick-Onyszkiewicz; for Joan Firey and Margaret Young. I cried for my father's family lost in the Holocaust and for my mother and grandparents in Chicago. I cried for myself, the little girl who wasn't allowed to go to the Lab School or play the piano. I cried for the end of my time in the glorious but truly dangerous world of the highest Himalaya, the exquisite icy world that I would enter no more.

I took a deep breath and returned to watch the dancers in the moonlight. We had achieved our goal. After Denali, Annapurna, and Bhrigupanth, there was no doubt that women could climb the highest and toughest mountains, both Himalayan and metaphorical. As my career as a high-altitude mountaineer came to an end, I felt content with my new dream. I would walk the Great Himalaya Range from end to end.

CHAPTER 21 ▲

My Father's Story

Chicago 1960

Sunday afternoon, my father and I go for a walk along Lake Michigan. He tells me about his childhood in Darmstadt, Germany, studying biology in a boys' school and vacationing in the Alps with his parents. In a quieter voice, he says his mother died in the 1919 flu epidemic when he was eight, and he grew up with his father and stepmother. He buys me a strawberry ice-cream cone and we sit on a bench, watching waves hit the shore.

"We had a good life in Germany when I was young," my father continues. "My father owned a large department store; we lived in a comfortable house and had many friends, Jews and Gentiles alike. But when I was studying at the university, Hitler took power. My father was very worried about German rearmament. Evenings he listened to Hitler on the radio ranting against the Jews. Gangs of Hitler Youth taunted me when I went out on the street. I didn't seem to have much of a future in Germany.

"When my father heard that Hitler was requiring Jews to get papers to leave Germany, he told me I should get out of the country until the storm blew over. In the autumn of 1933, my father hugged me good-bye and put me on a train for Paris, one that passed a small border post that didn't know that special documents were required for Jews to travel. I was twenty-two. I never saw my father again.

"From Paris, I made my way to Switzerland, where I went to medical school and climbed in the Alps during my vacations. My parents were planning to sell their store and join me, but they didn't sell it in time." My father closes his eyes and his head sinks down toward his chest.

I wait for him to speak, nervously twisting a strand of my hair. Finally I ask, "When did you come to America?"

"I came in 1938, after Kristallnacht. Hitler's storm troops destroyed the businesses of many Jews. My uncle Felix and aunt Yedel were living in New Jersey and invited me to join them.

"My father wrote me not to leave Europe. I felt guilty, but I came anyhow. I still don't feel so good about it. Then I moved to Chicago, where I met your mother." After another long silence, he continues.

"After the war, I went back to Darmstadt to talk with the people I had grown up with and find for myself what happened to my family. Our next-door neighbor told me that early in the war the Gestapo had come to take my father, who was nearly blind. My stepmother, who was much younger, insisted on going with him. They both died at Auschwitz."

I hear the waves splashing and children at play in the distance. I have been introduced to another grandfather and grandmother, and then they are both taken away. The story is more than I can bear and I push it out of my thoughts and watch the children.

After a few moments of silence, my father shakes his head as though to free himself of ghosts and we walk slowly back to Mitzi's.

We drink tea and eat Viennese pastries. Then my father asks me about my life. I proudly tell him I got all As on my last report card and am first in my class.

This is the moment I have dreamed of for years. I'd imagined the scene countless times. My father and I meet at last. I tell him of my accomplishments and the problems of life with my mother and grandparents: the loud TV, the smoking, the fighting, the haggling about money and food. He takes me in his arms and says, "It must have been terribly hard for you. But look how wonderfully you have turned out, all on your own. I am so proud of you and want to help you any way I can."

But instead he suggests that, after being raised by my mother, it might be useful for me to see a psychiatrist. He will pay for the cost of my therapy.

"No, thank you," I say coldly. "It was very kind of you to come and meet me." Not until I get into the car with Grandpa do I let myself cry.

The Great Himalayan Traverse, Part I ▲

1981

I FIRST MET HUGH SWIFT at a Berkeley party in the fall of 1980, attracted by the faraway look in his eyes and the round Himalayan cap on his head. After introducing himself, he mentioned that he was writing a book, *The Trekkers' Guide to the Himalaya and Karakorum.* Just back from India, I told him of our breathtaking trek across Ladakh.

As we traded stories of our Himalayan adventures, I felt as though Hugh were an old friend. Impulsively, I told him about my idea of walking across the Himalaya from Bhutan to Ladakh.

"That's the coolest trek I've ever heard of," Hugh said. "I'd love to do a trip like that."

I didn't know what to say. Should I invite someone I didn't know to share a yearlong trek? I hadn't yet worked out the details of the route or found my teammates; here was a guidebook writer, tall and good-looking, familiar with most of the areas I hoped to cross. Had fate thrown us together? He did seem a perfect traveling companion. On the other hand, I reminded myself, I'd invited climbers hastily for both the Endless Winter and Everest, and later wished I'd been more deliberate.*

"Listen," Hugh interrupted my musings, directing my attention to Bob Dylan rasping in the background: *To dance beneath the diamond sky, with one hand waving free . . .*

"That's my dream," Hugh said.

The spark I'd felt when I first noticed Hugh glowed more brightly. We agreed to meet the next day for a run. The connection between us was strong and immediate. During our run, just one day after we'd met, Hugh and I decided to spend the following year together, walking the entire length of the Great Himalayan Range. Only too readily I made my usual optimistic leap of faith that Hugh and I would be not only ideal trekking companions, but also possible life partners.

The Himalayan foothills are among the most beautiful, remote, and culturally rich regions on our planet, and at that time one of the

---

*It is not uncommon for climbers or other adventurous travelers who don't have regular partners to select companions quickly and sometimes unwisely out of necessity and optimism. For example, long-distance sailors frequently pick up crew members they barely know and head out to cross vast oceans. Sometimes a close friendship develops and other times the contrary.

few inhabited areas untouched by the modern world. We planned a route from east to west. Beginning near the eastern border of the kingdom of Bhutan, we would trek across Bhutan, the Indian state of Sikkim, the kingdom of Nepal, and finally the Indian Himalayan regions. Our destination was Ladakh, nearly three thousand mountainous miles to the northwest. As far as we knew, no one had ever completed this route. We called our venture the Great Himalayan Traverse.

As we prepared for the trip, Hugh and I shared details of our past lives. He was raised in a Quaker family and registered as a conscientious objector during the Vietnam War. Sent to South Vietnam to teach English, he was horrified at the American attacks and disillusioned by the deceitful way the war was represented by politicians and the media. After finishing his two years in Vietnam, Hugh traveled across Asia, stopping in Nepal, where he fell in love with the Himalaya and the people who inhabited its high valleys.

One evening, Hugh and I were in my little house in the Berkeley Hills tracing our route across several maps of west Nepal we'd spread out. Sitting together on the floor poring over the maps, we began finding each other as interesting as our proposed route down the Sano Bheri River and up the Thulo Bheri. In addition to a blossoming romance, Hugh and I discovered we had complementary abilities and worked well together. I would obtain the permits we needed and Hugh would choose our route. Neither of us could have done this trip alone and we were both grateful for the good fortune of having met when we did.

When I called the Bhutanese and Indian embassies for permits, I was told there was no precedent for our request, so no was the easiest answer. And that's what I heard, every time I asked for permission.

Needing some guidance on how to proceed, I contacted my friend Douglas Heck, a former U.S. ambassador to Nepal. He reminded me that in Asia, establishing a personal relationship is crucial. Following his advice, I again phoned the Indian embassy and this time introduced myself as Dr. Blum. Saying I would be speaking at the Smithsonian next month, I requested a meeting with the ambassador.

Three weeks later, I was in the ambassador's office, chatting and drinking cups of golden Darjeeling tea. At the end of the visit, with a dry mouth and racing heart, I asked the pivotal question. "A friend and I are planning to walk the length of the Great Himalayan Range," I said. "Our biggest problem is that foreigners aren't allowed to enter or leave India through the high mountain passes. Do you have any advice?"

"I will be most happy to assist you, Dr. Blum," the ambassador assured me. "I very much enjoyed our tea together."

And so had I.

On my way home, I stopped first in New York to see my father and then in Chicago to see my mother. I always hoped to find closeness and rapport during my family visits, but invariably left disappointed.

As usual, my father had a full weekend planned for us: a Tibetan art gallery in New Jersey Saturday afternoon, dinner at a kosher restaurant in the Bronx, a bakery in Queens for Black Forest cake, and an art gallery and Israeli folk dancing in Greenwich Village. I dutifully accompanied him, all the while tensed for his running critique: my dress was wrinkled; I ate too much; my manners were poor; his son Jonathan was much smarter than I was and his daughter Claire much nicer; I should get a job and take care of my mother instead of running around the world. My father was of the old German school that a parent's job is to criticize the child.

On the other hand, I heard from neighbors and friends that when I was not there he spoke of me with pride. He was charming to my friends, especially if they were attractive young women. On Sunday, he offered to show a friend and me Central Park, forgetting that the park was closed to cars on Sundays. Finding all the entrances blocked, he drove inside through an exit lane and proceeded to tour the park, scattering bicyclists and joggers in his wake. Hearing police sirens behind him, he accelerated and exited the park, muttering,

"What's all that noise? What's wrong with everyone today?" My friend and I clung tightly to our seats, disbelieving.

As I said good-bye to my father, he asked me to buy him Jewish menorahs in India, Nepal, and Bhutan.

My time with my mother was also challenging. In the decade since my grandfather had died, our relationship consisted mostly of my fending off her efforts to move in with me. Before each visit, I'd resolve to be patient and accept her limitations. But she was a strong, intelligent woman who continued to present her life as a desolate prison from which only I could free her. Sometimes she was so infuriating that in spite of my best intentions I'd end up screaming at her until my throat hurt. And then I'd feel terribly guilty. I tried my best to stay calm and feel compassion for my mother, but rarely succeeded.

After surviving visits with both my parents, I looked forward with renewed enthusiasm to the months of isolation the Great Himalayan Traverse would bring.

When I applied for a leave of absence from my job at UC Berkeley, I was turned down on the grounds that I'd taken too much time off for climbing already. Given the difficult choice between science and mountains, after some deliberation I chose to quit my job and head for the Himalaya. My decision was facilitated by Ronald Reagan's victory in the recent presidential election. Reagan was opposed to the regulation of toxic chemicals and it seemed unlikely that much progress would be made in my field during his presidency. The next four years seemed like a good time to take a break from my environmental research and focus on mountain adventures instead.

Of necessity, fund-raising had to be the first step of our trip. *National Geographic* magazine, which had supported our Annapurna climb, was a possible sponsor. Looking through my files one evening, Hugh came upon the old letter warning the magazine against publishing my Annapurna article. To my shock, Hugh told me that the return address was that of the noted Berkeley climber Galen Rowell. The woman who had signed the letter was Galen's girlfriend. Hugh and I both knew Galen and his partner and found it difficult to believe they would have written this slanderous epistle.

Berkeley being a small place, I happened to see Galen's girlfriend at a slide show the following week and I asked her about the letter. She paled and said she had not written it, but had known about it and had been extremely unhappy to have her name signed to it. She suggested we meet for a walk the next day to discuss the matter further. On our walk she told me the letter had been written by a group of Berkeley climbers, men who now knew they were wrong and deeply regretted what they had done. There was no point in telling me who they were. It would only cause me further pain.

I agreed, not wanting to revisit the anguish I'd felt from the letter. "Don't tell me the names now," I said. "Tell me when we're eighty. I don't want to think anymore about that letter or the people who wrote it. The letter is dead."

Hugh and I then spent six months organizing the trip. Friends and other trekkers would join us from time to time, and their payment for their short treks would help cover the costs of our long journey. L.L.Bean volunteered to donate all our gear and clothing, giving us a fashionably preppy look. After buying the toiletries and other personal items for the ten-month trek, we sorted and packed it all into ten large, heavy duffels. Then we made a somewhat sketchy plan of how someone would bring us a duffel each month.

We were heading toward a smooth departure in August 1981 when my preparations were stopped short by a cover line on the July issue of *Outside* magazine: HAS WOMEN'S CLIMBING FAILED? The article carried the subhead, "Why Has the Course of Women's Climbing Led to Tragedy?" It proclaimed in glossy detail that not only do women climbers tend to have inadequate skills and experience, they often climb for the wrong reason—to prove something about women. Building a case based on distortions and rumors, the author claimed that women who died while climbing high mountains were victims of their own pride and incompetence. His examples included Nanda Devi Unsoeld's demise on the mountain after which she was named, the Russian women who perished in the Pamirs, and Vera's and Alison's deaths on Annapurna. In the article, an "anonymous accomplished mountaineer" made the devastating accusation that Vera and Alison had died because I, trying to make a statement about women, had forbidden a Sherpa from accompanying them to the top.

I was shaking with indignation by the time I finished reading the article. Had the author even tried to verify his grave assertions? As any of the Annapurna team members would have told him, I'd first tried to persuade Vera and Alison not to make a second summit attempt, and then spent hours on the radio trying to convince the Sherpas to accompany them.

At the time, there was an unwritten code that when men die in the mountains, no one is blamed, even in cases of egregious negligence. Apparently the rules were different for women.

Irene Beardsley Miller and I wrote a detailed rebuttal, which we sent off to *Outside* magazine. Excerpts of our letter were published along with several other letters critical of the article.

The article had cited Galen Rowell as the source of allegations that several women had died because of their own mistakes. Galen had not actually been present on any of the expeditions he denigrated. Why did he judge so harshly these women who died climbing? I thought back to the letter to *National Geographic* signed by his girlfriend and sent from his address. Had he indeed written the letter? Was he also the "anonymous" source of the accusation that I'd caused Vera's and Alison's deaths? What was his problem with women climbing mountains? I decided to invite him for lunch and ask him directly.

I called Galen and he said he'd come the very next day. After a meal that neither of us ate, I reminded myself to breathe and asked Galen if he had written the letter to *National Geographic* and, if so, why it was signed by his girlfriend. Galen admitted he had typed the letter, but said the contents expressed the views of many other climbers; they told his girlfriend to sign it because they thought a woman's signature would carry more weight. He stood by the letter and the *Outside* article, insisting that our purporting to be a women's team while employing five male Sherpas was racist. On the other hand, our Everest expedition claiming to be American while employing forty Sherpas was a different matter. Emphatically, he said that women did not belong in high places unless they did it the "right way," which he didn't specify. I argued back, pointing out his lack of logic, but he neither backed down nor apologized for his anonymous accusations. After he left, I dictated our conversation into a tape recorder while it was fresh in my mind, entitling it "Lunch with Galen."

I couldn't understand his position. As a scientist, I looked for some rational explanation of his behavior. Why did Galen and his colleagues judge women so harshly and by such a different standard from that by which men were evaluated? No American men had yet climbed Annapurna. Did our accomplishment belittle their own climbs in some way? Or was it personal? Was there something about me that they objected to? My lunch with Galen had provoked more questions than it had answered.*

In late August 1981, after nearly a year of fund-raising and planning, Hugh and I left California for Bhutan, glad to be heading toward the clean air and long views of the high Himalaya. We arrived in steamy Delhi only to discover that we still needed an Inner Line permit from the Indian External Affairs Ministry before we could enter Bhutan and begin our trek. For a week we tried to get the essential permit, but all our entreaties were met with polite but firm refusals.

On Friday afternoon, the day before we needed to leave for Bhutan if we were to keep to our schedule, we sat hopelessly in the ministry waiting room. The office would close for the weekend in two hours. Without the permit, we would have to postpone our traverse or cancel the Bhutanese portion.

---

*In the years following these events, Galen often denigrated me to others in the outdoor community. Reports of these conversations came back to me from multiple sources. They were always painful to hear, and precluded my ever again trying to establish a collegial relationship with him. Galen's own climbing career was spectacular and his writings, lectures, and photographs were of the highest quality. Galen was a passionate and effective advocate for the environment and did much to call attention to the plight of the Tibetan people. Galen and his wife, Barbara, died tragically in a plane crash in August 2002.

Then I had an inspiration. We needed a personal connection, as with the Indian ambassador. Maybe we could find the person who actually issued the Inner Line permits. Quietly we left the lobby and wandered the corridors of the vast building, peering in office windows, not quite knowing what we were looking for. Then we found it: an office with a small man sitting at an ancient typewriter in front of a large green map of Bhutan. We knocked and went in.

Yes, he was in charge of typing the Inner Line permits needed to enter Bhutan, but we must know that these things take time. "At least six weeks," the man said. "You can get the forms at reception."

We explained that we needed our permit that very afternoon.

He looked startled and went back to his typing.

Hugh dug around in his backpack and pulled out a beautiful Himalayan calendar he'd had the foresight to bring along for just such an emergency. He gave it to the clerk.

"Thank you," the man said, looking at each of the striking mountain images. "Perhaps I can issue your permit sooner. Come back next Thursday."

I stared at the ancient ceiling fan going slowly round and round. "Please. Our whole trip depends on it," I said. "Is there any way you could issue the permit earlier?"

"Wednesday?" he offered.

"We need it today. Might that be possible?" I asked, holding my breath.

"I can have it ready for you Monday morning."

Hugh showed him newspaper articles about our Himalayan traverse and he offered us some tea. After tea, I asked again, "Can you please give us the permit now? Our flight leaves early tomorrow morning."

"Let me see your plane tickets," he replied.

He looked at them carefully. "These tickets cost many rupees. You must use these tickets. I will type your permit just now. Please wait. It will take only five minutes."

He typed the Inner Line permit, handed it to us with a slight bow, and we floated out of the External Affairs Ministry. We were the first Westerners to get permission to walk across the ancient Buddhist kingdom of Bhutan. Known as the Land of the Thunder Dragon, this small, landlocked country is located at the eastern end of the Himalaya. Most of the inhabitants are farmers and herders; nearly all are devout Buddhists.

A flight to Bagdogra, India, and a bone-jarring four-day drive two hundred miles east across Bhutan brought us to the starting point for our Great Himalayan Traverse. Tashi Yangtse Dzong, a temple fortress located near the eastern border of the country, housed both the local government and a monastery. Its high whitewashed earthen walls were topped with a golden pagoda-shaped roof adorned with dragon heads.

On September 20, 1981, Hugh and I, along with our staff of five Bhutanese—two guides, a cook, a horseman, and a Buddhist monk who came along to look after our spiritual well-being—headed west and up toward the Donga La, our first pass, at 12,500 feet. The next afternoon, Hugh and I hugged each other happily on the pass. Clouds were closing in, and I lingered on the descent, hoping for a view of the unexplored peaks to the north. Our guide had told me we would camp in a nearby meadow, but when I got there, it was deserted and I had no idea where the others had gone. A yak herder motioned me down a steep and muddy trail leading into a dense forest.

Three hours later I was still struggling down one of the worst trails I'd ever negotiated, a steep V-shaped trough of yellow mud filled with half a foot of water. Bloodsucking leeches lurked on every leaf. Here I was, on our second day out, lost, exhausted, and alone except for the leeches. Where was Hugh?

As darkness filled the woods, I heard the voice of our horseman, who had returned with a flashlight to show me the way. Following him down for an hour, I came upon Hugh at our camp at a small village *gompa,* or temple. He was sitting in front of a fire, eating dinner, his eyes glazed, the smell of hashish in the air. Sitting down next to him, I asked why he'd gone ahead without me. He told me not to worry so much.

The fact that he was high was not the issue. Climbers might get

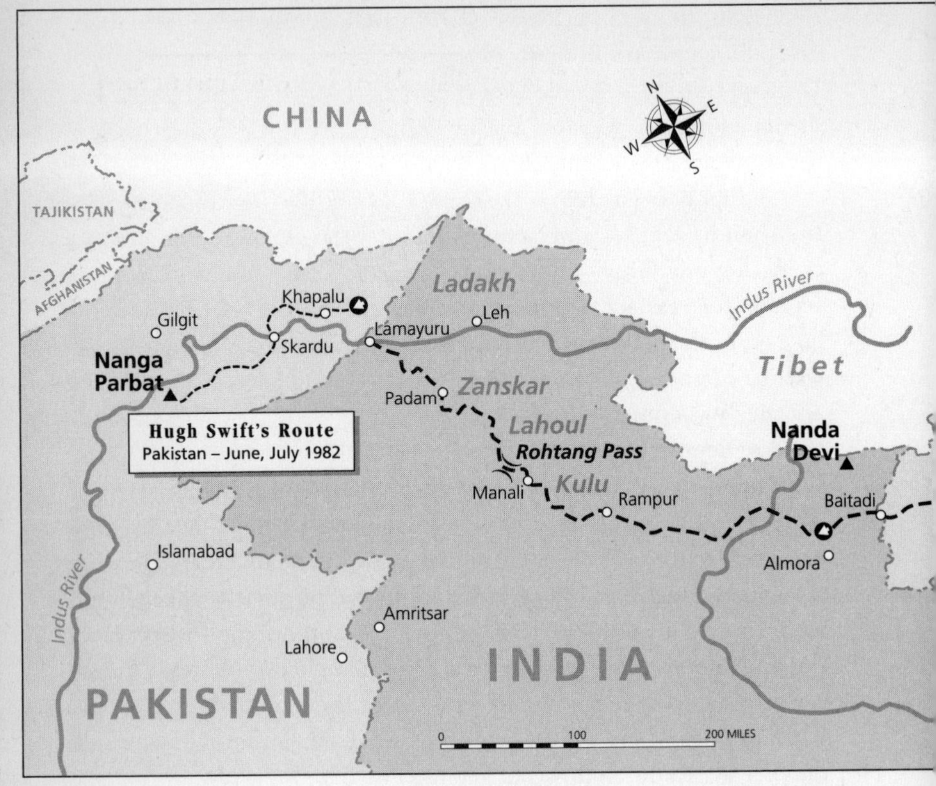

stoned, but would not leave a companion in the dark in a remote area. Ultimately climbers are there for their teammates, because climbing is a life-or-death activity. But Hugh was a trekker, not a climber. I realized with a sinking heart there was a lot I didn't know about this man who would be my partner for the next year. Had I done it again, chosen a companion too impetuously? Too tired to eat dinner or talk further to Hugh, I stumbled off to find my sleeping bag.

Our accommodations, on the wooden floor of the *gompa,* seemed luxurious after the horrific, leech-infested trail. A bronze Buddha gazed serenely from above a row of brightly burning butter lamps. Local people wandered in to say a few prayers, spin the prayer wheel,

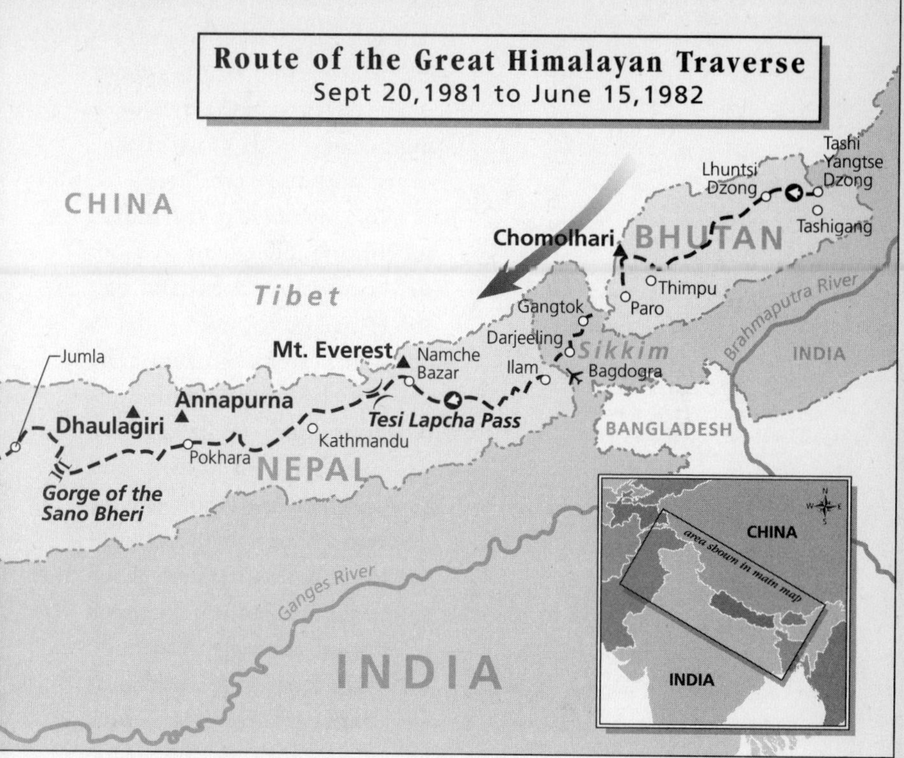

and stare in openmouthed wonder at us, the first Westerners they had ever seen.

Awakening to the quiet chanting of monks and the morning sun streaming through Buddhist prayer flags, I decided to forget having been left on the leech trail, and assumed everything was going to be all right with Hugh. He himself seemed unaware of any problem between us.

Just before lunch we reached Lhuntsi Dzong, where a major archery tournament was in progress. Each man lifted his three-foot bow, unloosed an arrow with a lunge and a great cry, and then ran after the arrow, yelling as if to urge it to the target—a foot in diameter and 130 yards away across a ravine. On the rare occasions when

A young monk carries water into Tashi Yangtse Dzong in Bhutan on September 20, 1981, as we begin our traverse.

the target was hit, the archers danced, sang, and swigged *arrah,* a fermented rice beverage. As the day progressed, the shooting became both more enthusiastic and more erratic.

The big rivers of the Indian subcontinent originate in Tibet to the north and cut through the Himalaya to reach India in the south. Trekking against the grain of the land, we would repeatedly climb up to a ridge or pass between 6,000 and 19,000 feet and then plunge down to a river, usually between 1,000 and 4,000 feet. Each time, the vegetation changed from tropical jungle to bamboo forests to temperate woodlands to mountain fir forests to rock and snow above the timberline and then back again. Some pass crossings were like a round-trip walk between the equator and the North Pole.

We frequently came to a village when a festival of some sort was in progress and were told, "You have come on a most auspicious day." Westerners had never before passed through these regions, and we were welcomed with extraordinary curiosity and courtesy. On one occasion we arrived at a small hamlet on the day of its annual masked dance and a young woman asked me if we were emissaries from the gods.

Crossing a 16,000-foot pass—our highest elevation in Bhutan—we reached a valley inhabited by nomadic yak herders and rare blue sheep known as bharal. Chomolhari, a 24,000-foot holy mountain on the border of Bhutan and Tibet, loomed above us. A family looking after three hundred large, hairy yaks invited us inside their octagonal yak-wool tent and we sat with them on a bearskin in front of a smoky fire beneath drying hunks of yak cheese. A mother nursing her eight-day-old baby chewed hunks of yak butter and then fed them to the infant.

A young Bhutanese shepherd welcomes us.

After dinner on our last evening in Bhutan, Hugh went outside for his habitual smoke. During our first month together, Hugh and I had grown no closer. In Berkeley, I hadn't paid much attention to his habit of self-medicating; it had seemed a harmless indulgence. But with the two of us so isolated, it seemed worth discussing. I summoned my courage and followed Hugh.

I sat down next to him and told him how much I cared for him, but that it was difficult to share things with him when he was on his own trip. Hugh was quiet for a minute and then told me that when he was in Vietnam he'd gotten into the habit of smoking hash to mute the horror around him, and that he'd continued this habit during his many solo treks in Asia. As he told me more about his experience of the cruelty and deceit of the Vietnam War, I felt enormous sympathy for Hugh. He then apologized and said he would try to be more present in the future. I felt our discussion had gone well and assumed the problem was solved.

We arrived in Gangtok, the capital of the Indian state of Sikkim, on October 21, 1981, surprised to see welcome signs throughout the town. We soon found out that the signs were for the Dalai Lama, the spiritual leader of Tibetan Buddhists, who was visiting here for the first time since 1958. Joining a colorful throng of Buddhists from all over Sikkim, we listened to his insightful teaching about peace and compassion. It seemed an extraordinary coincidence that the Dalai Lama gave an eloquent talk in English on our one day in Gangtok. Once again, we had come on a most auspicious day.

After a short stay in Sikkim we crossed into the Hindu kingdom of Nepal on the Day of the Dog, part of a Hindu holiday called Tihar. On this day, all the local dogs—even the mangiest street pariahs—were given a good meal and decked out with red *tika* marks on their foreheads and garlands of marigolds around their necks. The Day of the Dog had been preceded by the Day of the Crow and was followed by the Day of the Cow, the Day of the Family Money Box, and the Day of the Brother. During this holiday season villagers swung on high bamboo Ferris wheels and swings, believing that their upward motion would help the annual rice crop grow high and healthy. Because there was no electricity to power the Ferris wheels, a volunteer jumped up, grabbed the leg of one of the occupants, and pulled down to keep the wheel moving.

After two months, we reached the geographical high point of our journey: the bleak, windswept Tesi Lapcha Pass at 19,000 feet. For this ascent Hugh had reluctantly substituted an ice axe for the umbrella he habitually carried to protect himself against afflictions ranging from sun to rain to homesickness. To the north a monolithic rock peak jutted into the cobalt blue sky; to the south was Pacharmo, a huge, snow-covered mountain. I had crossed this same pass ten years earlier as part of the Endless Winter. I was happy now to be trekking the Himalaya instead of climbing them, to be enjoying the people and the beauty around me without the fear that someone might die.

On the crest of the Tesi Lapcha, Hugh and I looked back at the passes and valleys we had crossed since east Bhutan and forward to those that would lead us into India. We hugged warmly, for the moment content with each other and the reality of our dream all around us.

Chapter 22 ▲

My Grandfather's Story: A Barrel of Flour and Corn in Iowa

Chicago 1962

I need to get a summer job to make some spending money for college. I fill out applications at Jewel Foods, Tastee-Freez, and all the local restaurants, but no one wants to hire an inexperienced seventeen-year-old. After a sweltering day of rejections, I flop down by Grandpa on the porch. We sit in silence for some minutes and then I ask him to tell me my favorite story, about how he came to America.

"I grew up in a small village in Russia, near Kiev," he begins. "My father went around the countryside selling grains from his horse cart, but he wanted me to be an educated man, maybe a Talmudic scholar. Even though we were poor, my mother taught me to read and to play the violin. She was the only woman in our village who could read. I started to work with my father when I was much younger than you, but my mother made sure I practiced my violin and studied.

"My big brother Sam and I played violin, my brother Meyer played the horn. We all sang a lot on Shabbat." Grandpa smiles at the memory. "When Meyer was chosen to play the horn in Czar Nicholas's own band, we were very happy for him, but worried we'd never see him again."

"You all played music," I say. "I still don't get why you never let me."

"I've told you lots of time," he says. "Music was bad for your mother. She practiced so much it made her crazy. It wasn't so good for Sam, either."

This explanation still makes little sense to me, but I ask him what happened next, my favorite part of the story.

"In 1905, when I was sixteen, about your age, we heard that young men in our village were being taken away to be foot soldiers in the army of the czar. It was a death sentence. Everyone knew Jewish boys were cannon fodder. Once a boy left, he never returned.

"You know how your uncle Fieval has one glass eye? All he ever wanted was to study Talmud. To avoid being drafted, he poked the eye out on purpose."

I shudder and Grandpa stops talking for a moment.

"When we heard the Russian officers were coming, I climbed into an empty barrel. My sister Leah poured flour in the barrel, up to the top. I had a straw for air. I'll never forget how scared I was when I heard the soldiers enter the house and ask for me. My mother said I'd gone to study in Kraków, and then the officers rode away. My sister dug me out. I was white as a ghost, gasping, and covered with flour. We knew they would come back; we had to leave Russia.

"In 1906, our family left for America. It was so sad leaving our little village for a big country where we knew no one. For nearly a month we were crowded in the lower hold of a cargo boat with hundreds of other refugees. We were miserably seasick. The food was bad, the water worse.

"When we finally reached Ellis Island, the refugee worker asked my father what he did in Russia. My father said he sold grain and the social worker advised him to go to Iowa, where there was lots of corn." Grandpa chuckles.

"My family followed his advice. We took a train west and got off at the first stop in Iowa—Davenport, on the west bank of the Mississippi River. My father became a grain merchant, just like in Russia. Later, he helped found the B'nai Emes Temple, one of the first synagogues in Iowa.

"I got a job in a cigar factory, where I worked fourteen hours a day, every day except Shabbat. Those years were hard. Iowa was so different from our shtetl. The worst thing, I never told you before, my big brother Sam couldn't get used to the life in Iowa. He had five children and couldn't support them. All he could do was play his violin. He was very sad and he died by his own hand after the move. Such a terrible thing. I never wanted you to know this. But you know now. It was for your own good we kept you from the violin."

I want to protest that this still makes no sense, but he is back to telling his story.

"After four years, I saved enough money to start my own dry-goods store. I charged low prices and kept my store open long hours.

"I was twenty-seven and ready to settle down when I met your grandmother. We got married in 1916. There weren't so many Jewish girls in Iowa in those days and I liked how outspoken she was."

Just then Grandma yells at us to come in. Dinner is ready.

The Great Himalayan Traverse, Part II ▲

1982

DECEMBER CAME AND with it a succession of trekking groups that included many of our friends. For the next five weeks, all our needs were met by a team of local professional guides, cooks, and porters. We followed ancient trade routes through rice terraces and rhododendron groves across the relatively prosperous countryside of central Nepal. Trekking in the shadow of the Annapurna range with good food, weather, and company was a welcome change for both Hugh and me.

In early January we said good-bye to the last group of trekkers. We were about to enter the wilds of west Nepal, the most isolated and impoverished area of our trip. For the next two months, until we reached the Indian border, we would be on our own, entirely out of

With Hugh on Poon Hill in central Nepal, 1982. Dhaulagiri rises in the background.

The Great Himalayan Traverse family. *(From left)* Danu Sherpa, Prem Singh Gurung, me, Hugh, Pasang Tamang, Dil Bahadur Rai, and Dumbar Bahadur Tamang. Nepal, 1982.

touch with the rest of the world. When we told a Nepali family with whom we were staying that we were heading into the far west, they were alarmed, telling us of bandits and people who barely spoke Nepalese and practiced peculiar religions ahead.

To carry our loads, we had arranged to hire three porters, Dil Bahadur Rai, Danu Sherpa, and Pasang Tamang (the last names, Rai, Sherpa, and Tamang, denote the tribe or clan to which each belonged). Hearing that food might be difficult to obtain in the west, at the last minute we also hired Dumbar Bahadur Tamang and Prem Singh Gurung, porters from our last trekking group, to carry additional provisions. On an hour's notice, Dumbar and Prem agreed to spend the next six months with us, trekking all the way across west Nepal and India. I was very grateful for our porters' company, especially when Hugh was in his own world.

Most days I awoke early, full of good spirits. Being a slower walker than Hugh, I'd leave camp first and head up the trail, enjoying the wilderness around me. When Hugh caught up, I'd try to get a conversation going.

"Hi, Hugh. Isn't that an interesting bird? Is it a tree creeper, do you think?"

Sometimes we would start chatting and trek together companionably. Other times, Hugh would walk past me without a word and my mood would plummet. And then I'd vow to try to enjoy the trek by myself and not be bothered by Hugh's distance. Tomorrow things would be better, I was sure.

One sunny morning I got up early and hiked up a hill for a glimpse of the remote Kanjiroba peaks. I returned to a traditional Nepalese breakfast of rice and lentils, or *dal bhat,* and began to tell Hugh about my hike.

"It's amazing up there, flowers and a little lake and beautiful views of peaks."

Hugh continued eating his breakfast in silence.

"Let's take a day off. To walk up and get a better view of the Kanjirobas."

No response.

"Hugh, please answer me."

"I do *not* have to say anything to you," Hugh said in an icy tone.

Without thinking, I picked up my plate of *dal bhat* and threw it in his face.

Silently, Hugh stood up, rice and lentils running down his face and shirt, and strode off toward a stream.

The porters were horrified. To waste food was incomprehensible, as was to express anger. They hurriedly cleaned up and brought me a second breakfast.

By the time Hugh returned, a yellow stain on his shirt, I was repentant. I apologized and we did talk that day. But our differences were not easy to reconcile. Hugh wanted to walk at his own pace, explore and take in the experience in his own private way. I wanted a friend with whom to talk and share our adventure.

Our five porters were my best bets for companionship. I studied my language book diligently and practiced Nepali songs and conversations with the porters as we trekked along. My special friend was Prem Singh, an orphan of eighteen who was always singing, clowning, and telling funny stories. He seemed like a teenage son to me. Although the porters had never been in these areas before, they had an uncanny instinct for the correct direction, even in confusing snow-covered terrain. They told me they found trekking work easy compared with laboring back in their villages. Danu, our cook, said working for us was like an expense-account vacation with all the *dal bhat* they could eat.

And indeed, there were many fine days when the seven of us felt like a family, walking, dancing, and singing our way across the Himalaya.

Our focus was often finding food and a small, flat place on which to make camp. One dank January afternoon we went from house to house, looking for rice to buy. Unsuccessful and disheartened, we set up our tents. Suddenly a local Magar woman with an enormous nose ring appeared at the edge of our camp, bearing a pumpkin as a gift. Delighted, we offered to pay her for it but she refused our rupees. As the pumpkin, our first fresh vegetable in weeks, roasted in the fire, we were joined by a local shaman wearing a threadbare brown sports coat and large white turban. He presented us with a bag of "holy" rice.

"Eat this and you will succeed on your journey," he said. Hugh tried to pay him the going price. "No money," he insisted. "It is for the success of your trip."

I was touched by the hospitality these poor villagers extended to us "rich" foreigners walking through their land. Their generosity was extraordinary given the fact that the gear we carried represented more wealth than they could earn in a lifetime. Not to mention that we must have seemed to be on a perpetual vacation as we strolled past the fields where they labored to produce enough food to keep their families alive. I was painfully aware of our conspicuous privilege, but could find common ground when I played with the children, admired a baby, and shared a song or a polite greeting of *namaste*. As we walked along, local travelers would ask us what we were doing; the response that we were pilgrims would invariably elicit an understanding nod of the head.

To celebrate finding our way through the gorge of the Sano Bheri River, we decided to buy a goat. Danu was a Buddhist monk—and not eager to have goat on the menu.

I am touched by the villagers' hospitality in remote western Nepal.

"If you eat goat, it will snow," he told us solemnly. When he led a small, pert goat on a leash into camp, I was converted. "Let's not eat him," I said. "We can take him along as a porter to help carry our gear." However, now that the goat had been purchased, Danu thought differently.

"How many fleas have you killed on this trip?" he asked me.

"If they're in my tent, as many as I can," I answered.

"Then you shouldn't worry about killing a goat," Danu said. "Killing a flea or killing a goat is the same."

The next time I saw the goat, it was well roasted. Prem made a delicious dessert of fried dough for our feast. All that night, thick, wet snowflakes fell on our camp.

A few days later a local fisherman brought us his day's catch.

"If you eat fish, it will rain," Danu said.

We ate the fish, and the next day the deluge began.

Danu didn't warn us against eating chicken, but few were available in harsh west Nepal. One evening, we saw a plump chicken and offered the owner so many rupees that she sold us what was, apparently, her pet. She brought us the dead bird with tears in her eyes. And I cried with her, unable to eat a bite. Realizing that chickens produce eggs important for the diets of the poor villagers, I rallied to the non-chicken-eating cause.

Although we refrained from eating fresh goat or fish for the rest of the trip, the weather never improved. As we continued to march along through the sleet and storm, our porters teased me by singing, *"Memsahib maccha khanne bhane pani paryo,"* meaning: You ate fish; that's why it's raining!

Our goal now was Jumla, the regional capital of west Nepal, several weeks' walk away. The route was obscure and we weren't sure if we could even get there in winter. Repeatedly we climbed from a village at 6,000 feet where it was raining up to a 10,000-foot pass shrouded in thick snow. As we struggled toward Jumla, a fierce storm forced us to pitch our tents on the flat roofs of an impoverished village. When fleas and lice moved into our tent, I was desperate to move out, but the temperature was a chilling 33 degrees and the rain relentless. I contracted giardia and topped it off with a bad cold.

Those days in the flea-ridden tent, with everything around us cold and soggy, were the darkest hours of the trip. Hugh dealt with this adversity in silence. On the grounds that his side of our zip-together sleeping bags harbored more fleas, I took our bags apart—and as a vivid metaphor for our relationship they were never zipped back together. At this point we were many miles from the nearest road. We had no choice but to continue living close together in our small tent in spite of our strained relationship.

The storm finally ended and our trail headed straight up 1,500 feet through deep snow over rock. Along with an elderly Brahmin priest from Jumla and his teenage son, I kicked my way up a gully of deep powder snow. All of a sudden the old man sat down.

"What's the matter?" I asked.

"Go on. I'm going to die," he said. "I haven't eaten in three days."

I knew he was carrying bags of rice in his load and asked why he hadn't eaten. He explained that a Brahmin must wash in running water before he can eat rice, a sacred food. Because of the snow, he had been unable to wash and hence eat for three days. He'd run out of energy and could go no farther.

I reached into my pack for an enormous sugary Kendal mint bar that I kept for emergencies and gave him a piece, urging him to eat. He hesitated, ate a small corner of the bar, and then devoured the entire pound of mint candy. (Sweets can be eaten without washing.) I took his basket containing forty pounds of rice and iron cooking pots. Unburdened and energized by the sugar, the old man began to stride toward the pass. I could barely keep up with him while carrying his heavy load. Hugh, who'd been waiting for us at the pass, came back down to help. At the pass, the Brahmin thanked Hugh profusely for saving his life. Hugh and I looked at each other and laughed.

Sliding down the snow on the other side of the pass to green spring fields was easy. Then the Brahmin found a local well, bathed, and proceeded to cook a big meal of rice and dal.

Our last major obstacle before Jumla was the highest and toughest pass of all. The local people warned us it could not be crossed in winter, telling us of a villager who'd recently died in the attempt. Turning back and retracing our steps would preclude our finishing the traverse before the monsoons began in July, so we decided to try it. The snow was heavy and wet and no trail was visible. Just as we were thinking we might have to give up, four yaks came up behind us and, with no herder in sight, plodded steadily up toward the pass. The trail they broke was like a freeway, and we followed our four-footed guardians all the way to Jumla.

In early March, after a peaceful month of navigating our way across the rest of Nepal on a snow-free, low-elevation route, we crossed the border into India. Now we were no longer so isolated. Roads snaked their way up from the plains into the mountain regions, and villages near the roads boasted electricity, running water, and well-stocked

As we are about to turn back, four yaks mysteriously appear and break the trail leading us over the last pass before Jumla.

shops. I was dismayed when I saw my first car in months. Our days of being lost in time were over.

However, the rhythm of life in the Indian hills was similar to that in Nepal: brightly clad women returning from the forest laden with enormous baskets of fodder; men sitting around chatting and smoking hookahs; and barefoot children tending buffalo, cows, goats, and sheep.

Always, it seemed, it was the women who did almost all the work. At dawn they'd be high in the trees, lopping off branches that would feed the goats and cows, chopping and carrying the wood that would cook their food. They hauled the water for drinking and bathing, prepared the meals, took care of their homes and children, washed the dishes and clothes, and did most of the farm labor. They broke the soil with picks, planted, weeded, thinned, harvested, threshed, and ground the grain. We could hear them laughing and singing while they worked, and when we walked by they waved and called out to us.

In April we passed near Almora, where I spent two days taping an oral history of Gertrude Emerson Sen, a remarkable ninety-two-year-old American woman who had lived in the Indian Himalaya since the 1930s. She told me about meeting her husband, Basiwar Sen, who founded the Vivekenanda Laboratory for Hill Agriculture; their work with Mahatma Gandhi for India's freedom; and helping to

Hill women, young and old, work in the fields in the Nepalese and Indian Himalaya, 1982.

found the Society of Women Geographers (SWG) in 1925. I had first heard of Gertrude when I joined SWG, a network of women scientists, researchers, and explorers who have traveled extensively in little-known places, made original contributions in their fields, and added to the world's knowledge in a written, artistic, or photographic form.*

Continuing beyond a high village in the region of Rampur, our trail disappeared into wild, steep terrain. Hugh pulled out the map, which showed a blank area without a hint of a trail crossing the next pass. We realized that because of the roads, local people had stopped using the trails, which were now completely overgrown. Hugh and I alternated leading, fighting our way through brush and bamboo, balancing up and down on fallen logs. When we got hit by a sudden snowstorm, we had no idea which way to go. Then we saw a huge pink rhododendron tree in full bloom at the top of a steep, snowy slope. Energized by this incongruous vision, I kicked steps up the 30-degree slope to reach the tree. To our total surprise and delight we found the very pass we were looking for. Like our guardian yaks, this beautiful rhododendron had guided us in the right direction.

On the other side of the pass, we descended straight down a rotten cliff of rock, dirt, and bamboo, giving our porters a belay with our climbing rope. Finding a gully of soft snow, we giddily whooshed down, skiing on our boot soles, and Hugh used his open umbrella as a sort of parachute. Pitching our tents at the bottom during a hailstorm, we collapsed inside, grateful for a level campsite and a chance to rest.

The next morning, even our porters had no idea where we should go. As we dawdled over breakfast, eight men dressed in fancy jackets, pressed pants, and colorful woven scarves came sliding down the gully on the soles of their shoes. We were astounded to see them,

---

*In 1984, I was honored when the Society of Women Geographers awarded me their gold medal for "leading the first women's climbs of Denali, Bhrigupanth, and Annapurna." The medal had been presented to seven other women before me, including Amelia Earhart, Margaret Mead, and Mary Leakey.

as the people in the last village had told us that no one ever crossed the pass.

They were on their way to a wedding, they explained, and invited us to follow them down to Kulu. The only time anyone came this way was when there was a marriage between residents of the two high villages on either side of the pass. Once again we were in the right place on a most fortuitous day.

In the Kulu region the special days continued. It was the season for *melas,* or festivals in honor of the local gods. In Manali we enjoyed the biggest *mela* of all; seven heavy stone gods ornamented in gold and silver were carried in from seven villages to be honored by hundreds of dancers and thousands of spectators.

As we prepared to cross the Rohtang Pass into Lahoul, my good friend Anna Ferro-Luzzi came from Italy to join us, as we had prearranged. I moved out of the tent Hugh and I had shared for the past eight months and into Anna's.

There were plenty of tempests in our team after Anna arrived. An attractive and fiery professor of nutrition in Rome, Anna had little patience with Hugh's aloofness. The two of them had strong, opposing opinions on most things. As we trekked across the arid Tibetan plateau of Ladakh, I was back in my familiar role of peacemaker,

We arrive at Bardun Gompa in Zanskar, India, just in time for the dance of the four skeletons, part of the Buddha's birthday celebration, 1982.

Phuktal Gompa is carved out of the cliff face. Zanskar, India, 1982.

bringing food to Anna when she stormed off without lunch, asking Hugh to please wait for us at a good campsite.

After watching my attempts to placate Hugh, Anna told me that I seemed weaker and less decisive than before the trip. I was surprised, but I knew there was some truth to what she said. From my perspective, however, compromise was imperative if Hugh and I were to finish our Great Himalayan Traverse.

As we talked, Anna and I hiked through a dry valley between high, contorted rocky walls. I looked up and was struck by the hardy Himalayan pines clinging to the cliff faces above. Year after year these trees endured the harshest weather with little soil or water to sustain them, bending with the fierce gales but bouncing back when conditions improved. I felt a kinship with the sturdy trees. I told Anna that they were like me, raised with intermittent nourishment and many storms. It was comforting to see some benefits from my difficult childhood.

On June 15, 1982, Hugh, Anna, our porters, and I reached Lamayuru, a monastery situated in Ladakh on a barren hill just south of the India–Pakistan cease-fire line. When Hugh and I had begun from Tashi Yangtse Dzong in eastern Bhutan, nine months and roughly

The monks greet Hugh and me with prayer scarves as we complete our Great Himalayan Traverse on June 15, 1982, at Lamayuru monastery in Ladakh, India.

three thousand hilly miles ago, we'd had no idea where or when we would have to end our trip. Now we could go no farther.

Suddenly a monk approached us and asked, "Are you Hugh and Arlene?" In the mysterious way of the East, word of our coming had preceded us. And in a fitting close to our trek, we bent our heads and the monk draped our shoulders with long white prayer scarves.

From Delhi, our porters happily returned to their homes in Nepal with their wages. The money, more than they could have eked out from years of farming, was enough for Danu to build a house, Pasang to start a small export business, and Dil Bahadur to provide substantial dowries for his three daughters. Meanwhile, Hugh would complete his personal Himalayan traverse by walking with his friend John Mock across Pakistan for a month from Baltistan's Saltoro

Valley to the base of Nanga Parbat. He urged me to come with them and, in spite of our emotional distance, I was tempted. At this point it was the rhythm of the journey that most appealed to me. I was not quite ready to give up the days where my total focus was discovering a route across a windswept pass, finding some food and a small, level place near water to camp. But I had to return home to consult on a miniseries NBC was producing based on my book, *Annapurna: A Woman's Place.* With a mixture of longing and relief, I waved good-bye to Hugh as his black steam train pulled out of the Delhi station.

Hugh and I had shared one of our lives' greatest adventures. For months we'd been together day and night, navigating some of the most remote terrain in the world. Before we'd left, we'd wondered if walking every day for nine months would be boring. But every single day had been full of different challenges and delights. Although I regretted that the experience had separated rather than united us, I knew a trip like the one we'd just finished could be devastating even for well-established relationships. Two seasoned New Zealand climbers who did a similar trek were barely speaking by the end. A team of Indian women doing an even longer Himalayan traverse in 1997 split up into two separate groups, sparks flying from their ice axes, halfway through their attempt. Little wonder that Hugh, the solo trekker, and I, the extrovert, had such a hard time.*

Looking back, I sometimes wondered if the closeness and intensity of our trek was more than Hugh wanted or could handle. Perhaps he had no choice but to distance himself from me. I remember Hugh

---

*Nine years later, in February 1991, Hugh fainted while standing on a street waiting for a ride home following minor surgery in the San Francisco Bay Area. His head hit the sidewalk in a vulnerable place, and he died the next day. Like so many of Hugh's friends, I was heartbroken and disbelieving that this consummate trekker, a man who had walked more than fifteen thousand miles through the Himalaya without injury, could have died in such an unlikely manner.

Hugh's friend Eric Hansen later wrote *The Traveler: An American Odyssey in the Himalayas,* an eloquent tribute to Hugh's life of adventure. Reading it, I saw a different Hugh. Here he was a rock-and-roll pilgrim, passionately in love with the Himalayan people and land, a warm, insightful, and sensitive traveler. Trying to reconcile this image with the reserved man with whom I'd walked, I read a passage stating that later in his life Hugh longed for "the very things he had spent most of his life avoiding—intimacy and a sense of place."

as a unique individual with a great empathy with the Himalayan people, walking on his own terms through the high valleys he loved, carrying his umbrella.

Indira Gandhi, then prime minister of India, was a patron of Indian women's climbing and had heard about the Bhrigupanth climb and our Himalayan Traverse. After Hugh left for Pakistan and Anna for Rome, I was thrilled to receive a note from her inviting me to tea. Mrs. Gandhi wrote: "I have a deep love of mountains. One of the things I most miss because of my office is the ability to go trekking. I have never done real mountaineering myself, but I admire those who do." I was honored to have the opportunity to meet this modest woman who was governing India with such strength.

While we shared mild Assam tea, I told her stories of our adventures in her country. And then I brought up the questions that had been at the back of my mind for months.

"Why do the hill women do nearly all the work?" I asked. "Are they really as happy as they seem?"

I meet with Indira Gandhi after the traverse in 1982.

"Women have a high status in our society," Mrs. Gandhi answered. "They are responsible for the survival of their families. Their essential work gives them pride and self-respect."

"What about the men? I noticed them mostly sitting around drinking tea, smoking, and playing cards."

"The men do the plowing," she said. "And of course, they always take care of the money. Women can't do that."

"Why not?"

"Because they don't go to school."

"Why not?" I asked, having also wondered about that. Most of the schools we'd walked by had only one or two girl students amid dozens of boys.

"There is too much work for them to do," she said. "Girls don't have time to go to school."

There it was, the vicious cycle. Although the hill women did most of the essential work of growing crops, obtaining wood and water, and taking care of their home and children, they were denied education, legal rights, and control over money. Fathers, brothers, and husbands dominated them. Patriarchy exists in many societies—even wealthy ones where women have equal opportunities for an education and employment—but it is much more egregious in poor, agrarian countries.

Unsettled by our conversation, I politely folded my hands and said my farewell *namaste* to Mrs. Gandhi. Behind her, her turbanned Sikh bodyguards stood at attention. Two years later I would learn to my great sadness that two of her bodyguards had assassinated this formidable world leader.

With thoughts of returning to cross the Himalaya another time, I reluctantly boarded the jet that would take me back to my life in the United States. On the plane I took out a small book of sayings of the Indian Swami Vivekenanda that Gertrude Emerson Sen had given me for spiritual sustenance during our traverse. I opened it to a chapter titled "About Women" and was moved by the wisdom and current relevance of words the Swami had written nearly a century before.

In 1895, Swami Vivekenanda wrote: "The most important thing for the welfare of the world is to improve the status of women. A bird cannot fly on only one wing."

Chapter 23 ▲

My Parents' Story, Part I

Chicago 1962

The summer before I leave for college, I'm determined to find out what happened to my parents' marriage. I put together all the information I can glean from my family to construct my parents' story.

After college my mom and her sister Sylvia move to Chicago, get jobs at Spiegel's, and share an apartment in Hyde Park.

One autumn evening in 1942, my mother is watching folk dancing at the University of Chicago's International House when a tall, handsome young man with piercing blue eyes walks into the room. Although she notices him right away, she shyly refuses when he asks her to dance. He insists, takes her hand, and introduces himself as Dr. Ludwig Blum, a recent refugee from Germany. As they dance and talk they discover they are both Jewish, love music, and are lonely.

My mother is easily charmed by the cultured doctor with the exotic accent. My father sees a beautiful, proud young woman with a strong, square face and a clear gaze. When he tells her of his escape from Nazi Germany and his fears for his parents, her heart opens to him. Raised in small-town Iowa, she has little grasp of the maelstrom of World War II Europe or of the guilt that tortures him for having left Europe before the war. They begin to see each other every day. My mother, still unmarried at the advanced age of twenty-five, is delighted at my father's attentions and for the first time in her life falls in love. She ignores his moodiness and his critical nature, assuming their love will overcome all obstacles.

A few weeks later my mother invites my father to Davenport, Iowa, to meet her family. They spend a weekend with her parents and three sisters in the comfortable house where my mother grew up. Perhaps my father hopes to become part of this large, prosperous family and replace some of what he has lost. However, my father's German manners and somewhat imperious habits make my mother's parents uncomfortable and they take an immediate dislike to him.

My father asks my mother on a New Year's Eve date that will last all night. My mother says she will accept his invitation only if they get married first. On December 31, 1942, two months after that first dance, my parents are married by a justice of the peace.

My grandparents are devastated by this precipitous marriage and blame my father. Ashamed that they can't invite their friends and relatives to the wedding, they place articles in the local newspapers announcing the couple's engagement on New Year's Eve. Other articles in mid-January give a detailed account of a fictitious ceremony: "Miss Gertrude Isenberg and Dr. Ludwig L. Blum, whose engagement was announced recently, were married Sunday January 10 at 5 P.M. in a quiet ceremony performed in the home of the officiating rabbi Louis Lehfield. There were no attendants. The bride wore a wine colored velvet gown with black hat and accessories and her flowers were gardenias."

My parents set up housekeeping in a tiny Chicago apartment while my father looks for a surgical residency. Neither Germans nor Jews are welcomed by the American medical establishment in the 1940s, and he is turned down wherever he applies. For months he studies for the American medical boards while my mother works at Spiegel's to support them. Evenings and weekends, she does the housework, cooks, and tutors my father in English. Having been raised to play the violin rather than run a home, my mother struggles to meet my father's Germanic standards of housekeeping. Furious at the

My parents, Ludwig and Gertrude Blum, in 1943, shortly after their marriage.

man they believe stole their daughter, my grandparents are of no mind to help the newlyweds.

Frustrated by his inability to get a residency and anguished as he learns of the deaths of his family and friends in Germany, my father vents his emotions in angry outbursts at my mother. His moodiness and erratic temper frighten her and erode her fragile sense of self. As their marriage becomes increasingly troubled, they hope that having a child will bring them back together.

Coming Home ▲

1982

"SOMETHING TO DO, someplace to be, someone to love. That is the tripod we all need for a happy life," my Sherpa friend told me. "When one leg is missing, the tripod is not stable."

The sound of chanting monks drifted in the window as my wise friend and I drank tea in a small restaurant across from the huge stupa of Bodnath in Kathmandu. I was back in Nepal during December 1982, studying Nepali and trying to reconnect to the rhythm and beauty of my previous year walking across the Himalaya.

Before I had left for the Great Himalayan Traverse, I'd put my belongings in storage and rented out my apartment. Returning to California after that trek, I lived with various friends and tried to support myself by lecturing about Annapurna and our Himalayan adventure, all the while wishing I were back in Nepal. Since Hugh and I split up during the traverse, there'd been no romantic interest in my life. My friend's astute words about the requirements for happiness shed light on why I felt so disoriented. I wasn't sure if I wanted to live in Nepal or America, didn't know what I wanted to do, and had no one to love. All three legs of my tripod were wobbly, to say the least.

Unable to find a job in Nepal, I returned home to Berkeley. Remembering how happy I'd been when friends joined our Himalayan Traverse, I decided to organize and lead a series of educational treks. We would study photography, the Nepalese language, or Himalayan cultures while we trekked. I called my new venture Great Himalayan Treks.

At the same time, Charles Gay, Pam Ross, and I developed a lecture series called the Himalayan Trekking Course to introduce my clients and other travelers to the culture, politics, history, art, and geology of the Himalayan regions. The Center for South Asia Studies at UC Berkeley sponsored the course and we hung posters around town, hoping for at least thirty students. Nearly three hundred people showed up for the first class and we had to move to a large lecture hall. Believing that some knowledge of the language is a sign of respect for the local people, we taught Hindi, Nepali, and Tibetan in small sections before each lecture. Working with the language instructors, I developed a series of easy-to-use Himalayan language tapes for travelers.*

A Nepali language instructor at the Foreign Service Institute at the U.S. Department of State wrote me: "This course is especially good for people who have difficulty learning languages. Using these tapes, the students start speaking Nepali right away."

The concluding event of the class was a potluck in a Berkeley park. I photocopied Nepalese recipes for the three hundred people in our class and we had an enormous Himalayan feast complete with the students' hilarious attempts at Sherpa and Nepalese folk dancing. After the party Lama Kunga, our local reincarnate lama, suggested that we have another Himalayan celebration and invite the community.

His words came back to me as I strolled through a crafts fair in Live Oak Park a few weeks later. Surveying the booths of pottery and textile art, I imagined this ordinary event transformed into a Himalayan festival, just like the ones Hugh and I had encountered on all those auspicious days during our traverse. I envisioned tall poles with Tibetan flags blowing in the wind, sending up benevolent prayers. I imagined Himalayan food, dancing, arts and crafts, and the chaotic energy of an authentic Himalayan *mela.*

---

*Audiotape courses in Nepali, Hindi, or Tibetan for Trekkers and Travelers are available at http://www.arleneblum.com/language_tapes.html

Trekking to the Annapurna Sanctuary during my return to Nepal in 1983.

When I shared my idea with the next trekking class, several people offered to help plan a community Himalayan celebration for the following May. We prepared for half a year. The day before the fair, we strung brightly colored prayer flags high in the trees and set up the booths and food stalls. The next morning, the Nepalese vendors, Tibetan carpet merchants, and Indian dancers arrived, along with a few snake charmers and yogis. We waited anxiously. Would anyone else come to our fair?

During the next few hours, thousands of people streamed in. They shopped, watched the entertainment, ate the food, and had a wonderful time despite the fact we hadn't thought to provide toilets or trash bins. The fair made several thousand dollars, which we gave to charities and grassroots projects in the Himalayan regions, where a small amount of money can have a large impact.* The Himalayan Trekking Course, treks, language tapes, and fair combined to provide

---

*The annual Berkeley Himalayan Fair continues to celebrate the cultures of the Himalayan regions every year during the third or fourth weekend in May. Live Oak Park, located in Berkeley, California, is transformed into a Himalayan hill bazaar, and tens of thousands of dollars are raised for environmental groups, community development projects, women's groups, health clinics, orphanages, and schools in the Himalaya. In addition to being a joyous and lively community event, the fair gives us the opportunity to give something back to the mountain people. More information can be found at www.himalayanfair.net.

With the holy cow at the first Himalayan Fair, Berkeley, California, 1984.

me with a modest living and a community of people who loved the Himalaya as I did. Integrating my Himalayan life and my Berkeley life, I had found something to do.

In the summer of 1983, an Australian adventure-travel company invited me on a speaking tour across their country. I accepted, with the hope of finding a way to go scuba diving on the outer edge of the Great Barrier Reef, which I'd wanted to do ever since being trapped on Peak Lenin with *The Treasure of the Great Reef.* A Berkeley friend, Joan Seear, told me that her cousin Rob Gomersall had a diving business there and would be happy to show me the wonders of the reef.

Rob and I wrote to each other and arranged to meet after my lectures were finished. When I arrived in Rob's hometown of Airlie Beach, in northern Queensland, a friend of his met my plane. He told me Rob was off skippering the *Reef Encounter,* a large dive boat anchored in a sheltered lagoon eighty miles out to sea; I could fly out to join him first thing the next morning.

We left before dawn as crimson streaked the tropical sky, my tiny seaplane skimming above the nearby islands, the open ocean, and finally the Great Barrier Reef itself. The air was crystal clear and the blue and green lagoons shimmered alluringly. Off in the distance I could see Hardy Lagoon, where the *Reef Encounter* was anchored. We taxied across the water and stopped.

A motorboat driven by a fit, trimly bearded man made its way toward us. As he helped me into the small boat, he took my hand with a firm grip that lasted perhaps a few seconds longer than necessary. I looked into a handsome face and saw friendly blue eyes and a welcoming smile. The clarity of the day, Rob's direct gaze, his warm hand, our being in an ocean wilderness—all of it came together to make me feel that something extraordinary was happening.

An hour later I was floating in the warm water and then dropping down, down, down. Large gorgonian fans beckoned. Shining, silvery schools of mackerel and tuna hurried by. A striped cleaner fish nibbled parasites off a larger fish. An exquisite purple and gold nudibranch undulated past us. Rob led me through a complex of underwater passages from which we could pop back up to the surface. In the back of one cave, we came upon three huge purple and white lobsters in a circle, seemingly having a meeting. Exploring this perfect reef, with this attentive man pointing out wonders in every direction, I felt the same feeling of peace and homecoming I had when John Hall showed me my first glacier on Mt. Adams many years before.

At dusk Rob invited me on a night dive. Excited and slightly anxious, I slipped into the dark water. My light illuminated myriad floating shrimps and crustaceans, their eyes a shiny red. The fish were mostly asleep, floating calmly in rock crevices or even in the open water. A sea urchin nearly two feet in diameter pulsated iridescent green and purple. The coral, which had looked like stone during the day, had blossomed with colorful polyps reaching out hungrily for food. For nearly an hour Rob and I floated next to each other above a large coral head, watching the polyps feast on worms and small fish.

Motoring back to the big boat with a full moon hanging in the tropical sky, Rob told me that although the boat was crowded with tourists, he had a spacious, comfortable cabin all to himself that I was welcome to share. I felt like a character in a romantic movie. The strong British sea captain was inviting me to join him in his cabin. My next line was scripted. "Thank you very much. I'd be delighted."

For four idyllic days, we snorkeled, dove, and began to get to know each other. Too soon it was time for me to go on to Sydney and then my lecture tour in New Zealand. Rob and I talked about seeing each other again, but made no definite plans.

To my delight, Rob met me in Auckland at the end of my New Zealand tour two weeks later. Heading north to the Bay of Islands for a vacation by the sea, we went for a long walk along the rugged coast, talking about our past lives and future aspirations.

Rob told me he had a degree in ecology and had been a teacher in England. He had left to teach in Australia in his early twenties and made it his home. He had pioneered Land Rover routes across Tasmania and the interior of Australia and skippered boats worldwide. Now, at age thirty-seven, he worked as a dive instructor and a boat captain, and lived with a group of people in their twenties. "I feel like it's time for me to grow up and do something more substantial," he said. "But I don't know what."

"Close your eyes and imagine you could be doing anything you wanted, anywhere in the world," I suggested. "What would you do? Where would you be?"

Rob closed his eyes. "It doesn't matter what I do or where I am," he said. "All that matters is being with the woman I love." He opened his eyes and looked straight at me.

My heart lurched. Was he suggesting that I might be that woman?

"What do you think about my coming to visit you in the fall?" Rob asked. "I've never been to the U.S. and never especially wanted to go, but now I do."

Before I could respond, we saw a peculiar mound in the sand a few hundred feet ahead, gulls circling around it. Running up and shooing the gulls away, we came eye-to-eye with a massive dolphin that had been stranded in a small pool of water. The unhappy animal was lying on its side, miserable in the blazing sun. I filled my water bottle with cold ocean water and poured it over the dolphin. Somehow we had to keep the dolphin wet and cool until the tide came in again.

At a deserted sheep station above the beach I found a bucket, shovels, and a large tarp in an unlocked shed. Rob and I poured cool seawater over Debra (as we named the dolphin) until the tide washed in and she was in a foot of swirling ocean water. She seemed more comfortable, but the water was not deep enough for her to get back out to sea.

We put the tarp next to her and attempted to roll her over onto it. She cooperated, helping us move her enormous weight, more than five hundred pounds, onto the tarp. I felt her intelligence and love during that magical time as all three of us worked together.

Finally Rob and I dragged the tarp, with Debra on it, into the sea. She had been on her side a long time and seemed dizzy when we righted her on the tarp. We dropped the tarp and she rolled back over onto her side. We lifted the tarp up around her again, gently guided her into an upright position, and then slowly lowered the tarp. Again she rolled back on her side. On our third try, Debra hesitated for several minutes and then suddenly swam out to sea. Tears of joy ran down our faces. Rob and I sat arm in arm on the beach for an hour, but Debra didn't come back. We returned the things we'd borrowed from the sheep station and continued walking hand in hand along the beach.

Rob and I were in love. A week later I returned to California, counting the days until Rob's fall visit—and hoping we would be together for a very long time.

But Rob's life was busy and he decided not to come. When I received his letter I was inconsolable. Get a grip, I told myself, but for the next week I was in tears much of the time. Rob and I had been together for only a few weeks, and as powerful as our attraction was, I knew that my reaction to his change of plans was extreme. I decided to go into therapy to learn why I so often found myself sobbing as though the world had ended.

After I'd told Gerald Gray, my therapist, about my behavior, he asked me some questions about my early childhood, questions I couldn't answer.

At his suggestion, I called Aunt Ruth.

"I'm seeing a therapist this summer," I told her. "He asked me if Mom took care of me when I was a baby. Do you know if she did?"

"Honey, that was nearly forty years ago," my aunt said. "It's hard to remember. But my recollection is that after you were born, your mother was too tired to take care of you. She rested and we all fed you and did whatever you needed. You were well cared for."

"Did she take care of me later?" I asked.

"Not much," Ruth said. "As I recall, she had to be hospitalized shortly after you were born."

"Why?" I asked, trying to stay calm.

"She was depressed," my aunt said. "I think they gave her shock treatments, quite primitive in those days. It only made her worse, in my opinion. She's never been the same since."

"Why was she depressed?" I just managed to ask.

"Your mother could never conform to society," Ruth said. "In college they had to chase her out of the Music Department at one in the morning." My aunt paused. "Your father was so angry. He broke someone who was already not a strong person."

Putting down the receiver, I sank to the floor, feeling as though I were going to die. I felt a huge sorrow inside me, a black lump growing larger, threatening to burst and destroy me. I cried for all the losses in my life, for my mother's pain, and for myself as a baby and child separated from my mother.

Still crying, I called Gerald for an emergency appointment. The next day, I told him what my aunt had said and how I had reacted.

Gerald listened, thought a few minutes, and then said that my behavior made sense. "Very young children usually have a magical view of the world," he explained. "If a parent leaves, a young child may feel abandoned and that he or she is going to die. The terrified child reacts with uncontrollable crying."

I remembered the many times I'd found myself on the floor in tears: when Grandpa, John, and my other friends died, when Joel and Toby left me in Uganda during the Endless Winter, when John Percival wanted to spend an evening without me, when Hugh was in his own world on our Himalayan Traverse.

"That sounds right," I said. "Just like me whenever someone I care about leaves either physically or emotionally."

"Exactly," Gerald said. "Because your mother and father both left when you were very young, it's not surprising you feel abandoned when someone you care for leaves you. Those desperate feelings from early childhood are coming back and you are reacting as though you were still an infant.*

"Furthermore," Gerald continued, "the child can feel responsible and guilty for the parents' absence, as though the parents left because of something the child did."

"Like the way I manage to feel responsible whenever a friend dies in the mountains, no matter how unlikely the connection."

"Precisely," Gerald said.

These ideas were a revelation to me. Talking with Gerald over the next weeks, I came to understand that my mother's leaving for treatment of her emotional problems when I was very young and my father's leaving for good when I was three made me incredibly anxious about being left by anyone I came to love. Ironically, my desperate fear of abandonment tended to drive away the very people to whom I wanted to stay close.

As I began to understand myself and my emotions better, I was increasingly able to control my tearful outbursts. I now felt ready to

---

*Gerald read me an instructive passage from Selma Fraiberg's classic description of early childhood, *The Magic Years:* "For an infant, his mother is the link to the external world. Mothers and fathers, loved ones, are necessary for the child's existence and his inner harmony. When he loses them, even briefly, he is confused, disoriented, as if he had lost his connection with his new found world . . . when his mother returns, he has regained his world and found himself once more."

begin to share my life with a partner. I was thirty-eight years old and hoped it wasn't too late.

After that long summer of therapy came the Jewish High Holidays—Rosh Hashanah, the Jewish New Year, and Yom Kippur, the Day of Atonement. Years earlier I'd stormed out of synagogue, hurt and angry at the words that men recite each morning: "Dear Lord, thank you for not having created me a woman." Since then I had not attended services.

But that fall I wanted to return to the religion that had been the basis of our family life when I was a child. I decided to attend a Rosh Hashanah service sponsored by UC Berkeley Hillel at an unlikely location—the Catholic Student Center. Walking into the service, advertised as both traditional and egalitarian, I was astonished to see that the prayer leader was a woman wearing the tallith and yarmulke, the prayer shawl and skull cap traditionally worn only by men. Looking around, I saw many old friends from my years in Berkeley—climbers, scientists, and neighbors. Everyone began to sing and I felt the sound rise in great waves, filling me with a sense of belonging and unity with everyone in the room. As I joined the singing, tears streamed down my face. For golden moments, we stood arm in arm. Scattered among the congregation were a few old men davening—praying while moving back and forth rhythmically in the traditional fashion. Watching them, I let myself feel the pain of my grandfather's death twelve years earlier.

Seeing my tears, a friend hugged me and asked if I was all right.

"I haven't been this all right in a very long time," I reassured her.

At the service the next morning, I was asked to carry the Torah, the Five Books of Moses, handwritten on parchment and wound on two wooden sticks. For a woman to carry the Torah was forbidden in the Orthodox tradition in which I was raised. Trembling, I took the heavy Torah scroll and walked slowly around the room. Members of the congregation kissed the Torah as I carried it past them and I felt their love.

Even more revolutionary for me was hearing God referred to as "she" during half of the prayers. The idea of a loving, inclusive female God instead of the harsh patriarchal one of my childhood brought a new rush of tears.

The service alternated between traditional Hebrew prayers and discussions of world peace and compassion. In a true Berkeley service, we chanted prayers for Palestinians as well as for Israelis, for people of every nation, religion, and sexual orientation.

At the end of the service came the Mourner's Kaddish, the prayer for the dead. In all the years since his death, I had never gone back to a synagogue to pray for Grandpa. Almost trembling, I stood up with other mourners and chanted the ancient Aramaic words.

"Please say the name of anyone you'd like to remember and something about them," the prayer leader said after we sat down.

I wanted to recognize my grandfather but I couldn't speak. Everyone was naming people. The silence between names grew longer. I had to speak now or it would be too late. I inhaled and raised my hand. The prayer leader pointed to me. I exhaled.

"Isadore Isenberg, my beloved grandfather, who taught me the Hebrew prayers even though women were not allowed to read them in our Orthodox congregation." I had managed to publicly proclaim the gifts my grandfather had given me. I continued to cry, but now it was tears of happiness. I was accepting my past. My childhood needn't be a source of pain any longer. I was coming home at last.

Chapter 24 ▲

My Parents' Story, Part II

In February 1945, eight months pregnant, my mother leaves her job and her husband's anger and returns to her family home in Iowa. My father, now working at a public health clinic in Chicago, visits her some weekends.

I am born in Mercy Hospital in Davenport, Iowa, on March 1, 1945. My father finally gets a job as a doctor with the U.S. Army. My mother and I join him first in Chicago, then in New York City, Louisville, Memphis, and finally in a dilapidated converted army barrack in Shanks Village, south of New York City.

My mother and father are miserable—overwhelmed by their nomadic existence and their individual emotional problems. My father possibly finds comfort with other women. In his European culture having a mistress is acceptable; for my mother it is betrayal.

My mom and me on the front porch of the family home in Davenport, Iowa, as she brings me back from the hospital on March 11, 1945, ten days after I was born.

When my mother can no longer cope with her emotions and failing marriage, at my father's suggestion she voluntarily signs herself into a mental hospital. During the months she is away my father struggles to work and care for me.

When I am three my grandfather comes to visit us in Shanks Village. He finds his cherished daughter in a mental hospital, our home in chaos, and me hungry and crying with tonsillitis. He brings my mother and me back to Davenport, where my mother files for divorce on the grounds of emotional cruelty. The family views my father as a monster and my mother as a failure.

My mother and I live in the comfortable family home in Iowa for two years. My grandparents hope my mom will remarry, but there are few single Jewish men in Davenport.

When I am five, my mom is sent to Chicago to find a suitable husband for herself and a father for me. My grandparents and I go to Excelsior Springs, Missouri, where I attend kindergarten and they hope the hot springs will heal their various aches and pains. Perhaps they also are relieved not to have to explain their eldest daughter's disgrace to their friends.

Working again at Spiegel's in Chicago, my mother lives in a small room by herself. She doesn't go out on dates or make friends; she misses me that long, lonely year. Deeply depressed, she again enters a mental hospital, where she is given more shock treatments.

My grandparents are dreadfully worried about my mother and mortified by her illness. With the specter of the suicide of Grandpa's brother Sam in the background, they sell their large, beautiful home in Davenport, and leave their relatives, friends, and community. They settle in a small town house on the far South Side of Chicago, where they know no one. It is here that I spend the rest of my childhood.

Across the Alps with Baby ▲

1984–1987

IN THE FALL OF 1984, my focus was the presidential race between Ronald Reagan and Walter Mondale. Believing a second Reagan term would be catastrophic for working people and the global environment, I was doing voter-registration work full-time. Geraldine Ferraro was the Democratic vice-presidential nominee. My friend Terry Byrnes and I designed and marketed a T-shirt that read SUPPORT WOMEN'S CLIMB TO THE TOP—REGISTER AND VOTE. Jane Fonda modeled our shirts and thousands of people bought them.

A few weeks before the election, I received a letter from Rob Gomersall, who was now teaching sailing and navigation in Canberra, Australia. Even though I hadn't been thinking about Rob very much recently, I was nonetheless delighted to hear from him.

"I just bought some beautiful bush land, a perfect place to build myself a house," he wrote. "When I was there last week, I saw a nest with three young kookaburras in a hollow tree. The mother and father birds were feeding their babies with great care and love. Watching them, I felt there was little point in building my house alone. Would you like to come to Australia to see my land?"

I felt a rush of tenderness toward this kind man who wrote me about baby birds.

On my way to leading a Nepal trek in November, I stopped in Canberra to spend a couple weeks with Rob. After a sweet welcome, he showed me the wild and lovely land where he planned to build his house. The magic between us was as powerful as ever, and I accepted his invitation to return to Australia after my trek to spend Christmas with his parents in Melbourne. Then we would go back to America together.

With Rob Gomersall on his land near Canberra, Australia, 1984.

However, when Rob came to California that winter, he didn't like the climate, the culture, or the hectic pace of life in Berkeley. Similarly, I felt isolated and unhappy when I tried living with him in Australia. We were not willing to let our relationship go, however, so we decided to tackle head-on the difficult problem of bringing our lives together.

For a start, we merged our work lives by adding Rob's Australian diving business, Reef Promotions, to my Great Himalayan Treks. Commuting together among America, Nepal, and Australia, we marketed our trips in the United States, Rob assisted with my Nepal treks, and I helped out with his sailing and diving adventures on the Great Barrier Reef. At the same time, we struggled to make our relationship work after so many years of independence.

By the time I turned forty-one, we knew if we wanted to have children we would have to act soon. My own childhood hadn't been a good advertisement for family life. For decades I had been focused on science and mountains; motherhood was on the distant horizon. But as my confidence and self-awareness increased, I felt ready for parenthood.

Then, on a most auspicious day, I learned I was pregnant. Just in time for me, Rob and I were given the opportunity to become parents. We were very grateful.

My mother was worried about me during my pregnancy, perhaps recollecting her own travails during that time of her life. She had long had the disquieting habit of sending me heavily annotated newspaper clippings on every possible catastrophe, from tornadoes to child abuse. During my pregnancy, she sent these missives almost every day. She also sent me dolls, little purses, and other gifts appropriate for a five-year-old, each wrapped in layers of boxes and tissue papers with multiple ribbons. I managed to keep some perspective, thinking the gifts might at least be appropriate for my baby in a few years.

Rob and I were scheduled to lead a trek to Everest Base Camp during what would be the seventh month of my pregnancy. A friend had the temerity to suggest that leading a trek to 18,000 feet might not be the most appropriate activity for a forty-one-year-old mom-to-be. I consulted my trek-leader friend Charles Gay for a second opinion, and he agreed that Everest Base Camp was not an optimal third-trimester destination.

Instead, Charles had a most interesting proposal. After telling me that he and his family were moving to Nepal, he suggested that I buy their house near the top of the Berkeley Hills. Rob had sold his land and neither of us lived in a place suitable for raising a child, so the issue of where we would live with our new baby was becoming urgent. With little deliberation, I went ahead and bought their simple home, which was close to trails and open space. The house had found its way to us; I hoped that living in it, Rob would be happier in Berkeley.

My pregnancy was euphoric, but my labor was long and painful. Even with Rob's valiant support, after eighteen hours I'd had enough.

"You can do it," the doctor urged me on. "You've climbed all those amazing mountains. This is easy compared to that."

"Right," I retorted. "I'm leaving now to climb the mountains; you stay here and have the baby."

I was exhausted and our daughter showed no indication of wanting to be born. The doctor said there was a change in her vital signs and it was time for him to intervene. Though I had resolutely wanted natural childbirth, at this point, I was ready to take or do anything to end the ordeal. The doctor tried attaching a suction cup to her head, but it slipped off her thick hair. Finally he pulled her out with forceps. Our little miracle was an eight-and-a-half-pound perfect baby girl, with a large round face, my dark curly hair, and Rob's big blue eyes.

Annalise Gomersall Blum was born at 3:29 P.M., and we left the hospital at 3:28 A.M. My health insurance did not have good coverage for childbirth and there was a bargain rate if we stayed twelve hours or less at the hospital.

The next weeks with this precious new person were more miraculous to us than any mountaintop or deep-sea reef. Nevertheless, Rob and I weren't ready to settle down to a life of diapers and domesticity. Before long we were brainstorming adventures appropriate for baby Annalise. The following summer, when she would be four to six months old and portable, but not ambulatory, seemed a perfect time.

Rob suggested a sailing trip, but nursing and changing a baby aboard a rolling boat did not sound appealing. The trans-Siberian railway seemed monotonous; the possibility of getting sick in Asia was too great; Antarctica was too cold and Africa too warm. Because babies don't adapt well to high altitude, the Himalaya were out.

Finally we came up with a perfect place: the European Alps. They were a low-altitude, civilized mountain range boasting spectacular scenery, well-marked walking trails, and conveniently spaced huts providing food and shelter. We decided to hike the length of the Alps, carrying baby Annalise on our backs.

Suzi Le Bon, a Berkeley friend who had attempted a solo crossing of the Alps, suggested a route west from the Yugoslavian border across the alpine regions of Austria, Italy, and Switzerland, to Chamonix, France. Most important, Suzi assured us that disposable diapers would be available in the villages along the way. Estimating the distance to be about six hundred miles and six hundred diapers, we gave ourselves three months, hoping Annalise did not start actively crawling or painfully teething during that time.

Our concerns were slightly different from the usual worries of first-time parents: Would we be welcome in the huts? Would she be able to sleep soundly in so many different places? Would we be out in the cold if she cried in the night? Would my milk supply last with all that exertion? What if Annalise needed her diaper changed on a pass during a thunderstorm? Would we be polluters if we washed her bottom in a mountain stream? And finally, how could parents who worried so much about everything dare attempt such a trip?

We didn't have the answers to any of these questions in June 1987, when the three of us left Yvoire, near Geneva, on an overnight train bound for the Yugoslavian–Austrian border. A local taxi took us from the train station to the border post in the pelting rain. We got out, bundled Annalise into her baby backpack with rain fly, and looked with trepidation at our first trail. According to our map, eight miles of walking would bring us to the village of Thor Magnellan. Imagining a dark, dripping-wet slog along a muddy track and a wet, unhappy baby, we sheepishly asked the taxi driver to take us to the village.

We received a warm welcome at a cozy guesthouse, where a motherly Austrian woman ushered us to seats by the fire, gave us steaming mugs of rich, hot soup, and showered us with questions and praise for our beautiful baby. When we told her our plan she looked worried, got up, and riffled through a stack of newspapers. Beneath a headline proclaiming THE SUMMIT OF IRRESPONSIBILITY was a photo of a man carrying a toddler in a backpack. The front-page article reported that a Polish climber had tried to take his two-year-old son up Mont Blanc but had been stopped at the top hut by French police, who accused him of child abuse. The father and son were forcibly removed from the hut by helicopter, at the father's expense.

Great timing, I thought. People will be alert for mad foreigners carrying their babies up mountains. We assured our concerned hostess that we had planned a conservative route that would not go anywhere near the summits.

With baby Annalise in the Austrian Alps, 1987. *(Photo by Rob Gomersall)*

The next morning, the sun shone brightly as we took the local bus to the peaceful village of Achomitz, where we helped each other on with our enormous packs. As first-time parents with vivid imaginations, we were prepared for any emergency. My custom-made baby backpack bulged with a huge first-aid kit containing drugs for every imaginable pediatric illness, bottles and formula in case my milk ran out, disposable diapers, and cloth diapers and diaper covers in case we ran out of disposables. My pack weighed about sixty pounds and Rob's more than seventy. Perched on top, wearing her baby sunglasses, Annalise played contentedly with Dolly, Teddy, and an assortment of rattles and teething toys that dangled from the sunroof of my pack. Somehow we had managed to leave her high chair, car seat, and Swingomatic behind.

The tinkle of cowbells provided a soothing accompaniment to the walk to our first hut. When we arrived, the beaming, red-cheeked hostess served up thick slices of fresh-baked bread with creamy, rich butter and delicious homemade cheeses. She cooed over Annalise, the first baby ever to visit her hut.

After a substantial dinner, we moved into a small room under the eaves, complete with two kittens. Most hut accommodation is large dormitory-style, but our family was given the best private rooms available.

While Annalise slept, Rob and I began to discuss where and how to fashion our life together; so far, neither of us had been happy in the other's country. I was pleased when Rob suggested a creative compromise: New Zealand. I liked the people and mountains there, and thought that living near a university would allow me to connect with like-minded folks. Auckland or Christchurch would be perfect.

Rob thought otherwise. "Those cities sound as bad as Berkeley to me. How about a small farming town or an island?"

I did not want to be in a remote place in a new country with a new baby. But more than anything in the world, I wanted the three of us to be together as a family. There had to be a way. We let the

Annalise eats her first solid food in the Italian Alps, 1987.

conversation drop, assuming that we would have plenty of time to bring it to a satisfactory conclusion.

Mostly we stayed in the present, enjoying the flower-filled valleys, spectacular vistas, and comfortable huts that filled our days and nights. I nursed Annalise in meadows of brilliant blue gentians and columbines, yellow buttercups and pink primroses, on the edges of blue- and green-tinged glaciers, and on narrow trails carved out of limestone cliffs. I changed her diapers with backdrops of sheer granite spires and orange and pink mountain sunsets. We found water to wash her in gurgling mountain brooks and milky torrents pouring from the bottom of melting glaciers. She enjoyed her first solid food—freeze-dried peaches—on the edge of a tumbling glacier in the Italian Dolomites. Many days we moved laboriously up steep trails to high passes, disturbing the alpine peace with our loud, breathless rendition of "Old MacDonald Had a Farm." Our baby's name attracted considerable attention, too. German, Austrian, Dutch, Italian, Belgian, and French hikers all told us that Annalise was a charming traditional name from their country.

Not everyone was delighted at the sight of us. Some Austrian alpinists, decked out in heavy boots and carrying elaborate equipment, were most disapproving. "It is not possible to cross that pass with a baby," they admonished us sternly, glancing askance at our huge packs and jogging shoes. Our welcome was similarly unenthusiastic when we stayed in a Swiss hut whose other occupants were planning to get up at 1:00 A.M. to climb a difficult peak. I tried to warm the frigid atmosphere by explaining that Annalise would not disturb them. And she did not, kept quiet and content by nursing most of the night.

Annalise was healthy and happy all summer, and we never needed to use our baby first-aid kit. The admiration showered on her by the people we met and from both her adoring parents helped her develop into a cheerful and sociable infant.

In early August, when we reached about the halfway point of our walk, Rob received an urgent message that his mother had just had a serious stroke. He left for Australia immediately, and Annalise and I joined him soon afterward.

Rob's mother, Eileen Gomersall, passed away a month later. We were all very sad, especially because Annalise and her loving grandmother would never know each other. Overwhelmed with melancholy, Rob didn't want to talk about our future during our time in Australia. The window of opportunity for the major give-and-take necessary for us to find a life together seemed to be closing. In the Alps, high above the world with our sweet baby, we were both open to compromise. Back at sea level, however, finding a mutually agreeable plan was harder.

For several years we tried to make our relationship work, wanting to become a more traditional family. However, each of us had our own home and life half a planet apart: mine in Berkeley and Rob's in tropical Queensland. The next summer, Annalise and I came to live with Rob, but he was away for part of it, sailing with a group in a tall ship in New Guinea. Then Rob arrived in America with his mother's wedding ring to give me, but we still couldn't agree on where and how to live. We were always happy having adventures together, but regular life was harder. Although I longed to share the miracles of Annalise's growing up with Rob, sadly enough we could never find a way to build a nest and raise our baby bird together.

Motherhood was indeed my life's favorite adventure. In many ways it was the culmination of my experiences, calling for all the stamina and determination that had been honed in the mountains; the creativity and attention to detail necessary for scientific research; and the love I had longed to give and receive my entire life. My heart, mind, and soul were focused on Annalise, a being who filled me with joy.

Looking forward to new adventures, in the French Alps with Annalise, 1987.
(Photo by Peter Molnar)

CHAPTER 25 ▲

The Wheels on the Train
August 1962

The wheels on the Northwest Zephyr *steadily move me from my childhood home in Chicago to my new life in Portland. As the train speeds through the corn- and wheat fields of Illinois, Iowa, and Kansas, I am in free fall. I float, weightless and liberated, exhilarated at having separated from the powerful gravity field of my family and filled with anticipation at being pulled into my future at Reed College.*

I make my way to the dining car, where a grizzled waiter leads me to a table set with linen napkins, mirrorlike silver, and a red rose. He suggests the dinner special, filet mignon. My mouth waters as I realize I can order my very own steak. Looking at the prices, I order a tuna sandwich.

Eating slowly, I watch the golden glow of the late-afternoon sun blanket the wheat fields outside the dining-car window. Suddenly something white and unfamiliar appears far in the distance. The vague mass sharpens into a row of floating white pyramids. As we get closer, the vision turns into a range of huge mountains of snow and rock. I am filled with an intoxicating sense of possibility that only grows stronger as we move toward the Continental Divide. The flat homogeneity of the Midwest is behind me, the Rocky Mountains lie ahead. I want to stop time and hold on to this moment, but the train moves unceasingly westward.

Peace and Love at Last ▲

1993

ONE AFTERNOON AS I RAN on the fire trail above my house, I met up with Grant Barnes, a climbing friend and American Alpine Club member. As we jogged along chatting, our conversation turned to my difficult relationship with the climbing establishment. I'd never understood why I'd been ostracized for going to the Pamirs or why the AAC initially refused to endorse our Annapurna expedition with me as the leader. Years had moderated my hurt, but it still rankled.

Grant told me that as chairman of the AAC's Publications Committee at the time of the Annapurna climb, he'd heard the discussions questioning my leadership and had long meant to offer me his interpretation of what had happened. He believed there were several underlying reasons the climbing establishment had found me unacceptable. In addition to being a woman, I was Jewish, tall, outspoken, and I had a doctorate. We both laughed at how ridiculous it sounded, but a part of me felt like crying.

Grant went on to say that some outspoken members of the AAC board reflected the prevailing culture of that time—patriarchal, somewhat anti-Semitic, and not ready to deal with strong-minded women. To them, I was an alien creature. Proper women stayed home and took care of their families or else accompanied their husbands or boyfriends to the mountains; they didn't initiate or lead expeditions. We ran in silence as I took in Grant's words, which were a revelation and a relief. I'd taken a share of the responsibility for the discrimination I'd experienced, knowing that some of the things I had done, like going to the Pamirs without American sanction and organizing all-women expeditions, were incendiary for some climbers. But I'd also worried that some unknown character flaw of mine had contributed. So I was relieved to have Grant's perspective. I couldn't fault myself for my gender, religion, height, directness, education, or interest in leading expeditions.

"I *really* appreciate your sharing this with me," I said as Grant and I finished our run. "It will help put some old ghosts to rest."

Shortly afterward, six-year-old Annalise and I went to Chicago to visit my mother and my aunt Ruth. My whole life I had wanted a mother who would nurture me rather than one who wanted me to take care of her. Now, at age seventy-five, my mother seemed to have accepted her life in her pleasant Hyde Park apartment and was competently managing her own affairs. She told me about enjoying symphony concerts and playing her violin again. Annalise was studying violin, too. My mom listened to Annalise play and then played for her in return. Mom gave Annalise a gift; enchanted, my daughter unwrapped the layers of paper and ribbon around a small portrait of Mozart. As Annalise thanked my mother with a hug and a kiss, I was moved by the warm connection between them. Being a grandma seemed to suit my mother, and for a brief moment we seemed like three generations of a regular family.

Seeing my mom doing so well at last gave me a burst of confidence. After a cordial dinner with Aunt Ruth and Uncle Sy the next night, I got up my courage and asked Ruth about the conversation that had haunted me since I'd overheard it that August afternoon so many years before.

"Do you remember, when I was about four, you and Aunt Shirley were talking on the back porch in Davenport?" I asked, unable to keep my voice from shaking.

"Gosh, honey, that was so long ago," Ruth said.

"I heard Shirley tell you that I would amount to no good." I just managed to say the hurtful words out loud. More than forty years later, the memory still brought tears to my eyes.

Ruth looked surprised. "No, I don't remember anything of the kind," she said. "Shirley would never have said that."

"I'm sure she did," I said with certainty, no longer shaking. "I've never forgotten it."

"Maybe you didn't hear right."

"I can hear her words in my mind as if she'd said them today," I responded, looking Ruth straight in the eye.

"Well, maybe she said it, but I wouldn't take Shirley too seriously," Ruth said. "She can go on and on without thinking. She likes to stir the pot. I'm sure she didn't mean it. I'd just forget it if I were you."

Just forget it? Those words had troubled me for most of my life and Ruth was telling me not to take them seriously?

But thinking about what Aunt Ruth said, I was able to gain some perspective on Aunt Shirley's comments. Her words had served as a cornerstone of my sense of unworthiness, but they were far from the only thing that had made me feel so much an outsider during my childhood. Even at age four I didn't need Aunt Shirley to tell me that things were not as they should be in my family.

But now, half a century and countless experiences later, I knew that difficult as my early years may have been, they contributed to my strengths as an adult. And, in some unknown way, Aunt Ruth's words completed my healing. The words I'd overheard from under the porch finally lost their power. It was like awakening from a bad dream; I was left with the memory of having been very distressed, but the unpleasant emotions were gone and couldn't be conjured up again.

A few months later Annalise came east with me for a lecture trip. First we stopped for a brief visit with my father in New York City. He was loving with Annalise, but, as always, critical of my life and my parenting. I continued to hope for a meaningful connection with him, but could not get past his forbidding exterior. When we were apart, I could appreciate his creativity and imagination. But when we were together, I felt hurt by his emotional distance and judgments of me. To my continued sorrow, my relationship with my father never came to resemble that of my childhood dreams.

Then Annalise and I continued on to Boston to visit my cousin Mike, who lived with his wife and two young daughters in a large, beautiful house surrounded by classic New England woods. They offered us lunch when we arrived.

"Want some blueberries?" my cousin asked, and when I said yes he put a large bowl of berries in front of me. I looked over at Annalise, who was playing with her cousins, every bit as bright and happy as they. I smiled to myself. I had climbed high mountains, done important scientific research, and here in front of me were as many blueberries as I could eat.

I relished the taste and abundance of the succulent berries, the fruit I'd longed for as a child. I gave thanks for my beloved Annalise and the rich life we shared, our rustic home with a view of the San Francisco Bay, and my interesting work. It was neither the peaceful, solitary life of my childhood fantasies nor the big, happy family I had hoped for later, but the legs of my tripod were strong and stable.

Backpacking with Annalise in Garibaldi Park, British Columbia, 1991.

Epilogue ▲

Mountains, Molecules, and Motherhood ▲

This memoir ends in 1993, when Annalise was six and I had found some measure of peace with my family, myself, and the mountains I'd climbed and those I would never ascend.

I still enjoy climbing in the Sierra Nevada and other Western ranges; sometimes Annalise and I include a peak such as Mt. Kenya in the trekking trips we lead around the world. But I no longer choose to attempt summits of the life-threatening variety.

Selecting the photographs for this book from thousands of dramatic images of my expeditions was a daunting task.* Looking at these photos, I am reminded of the incredible beauty of the highest mountains and how privileged I am to have lived and climbed amid these magnificent surroundings.

At the same time, I will always be haunted by the high price that comes with this beauty.

When I visited John Hall's parents—still married after sixty years—in Portland during the spring of 2004, John's mother showed me the letters of condolence she had received after the avalanche on St. Elias. Still hoping to better understand my enigmatic first love and climbing mentor, I got in touch with several of the letter writers. We had lengthy conversations from which I received the impression that these other friends of John's, both women and men, were on some level still in love with him, just as I am.

I continue to share with Bruce Carson's and Hugh Swift's families the heartbreak of losing their wonderful sons. My own daughter, now a teenager, has no interest in high-altitude mountaineering, and I cannot say I'm sorry.

---

*Additional color photos documenting the adventures in this book can be found at www.arleneblum.com.

Currently, in 2005, only my mother and her youngest sister, Shirley, remain of the redoubtable Isenberg clan. Now that my mother is eighty-eight, I am happy to care for her as she had always wished. Gravely ill as I complete this book, she recently rallied to listen to my reading all the family scenarios to her over several days. The time was a gift for us both.

Sadly, my father and I never reached such a place of understanding. Our relationship remained difficult until his death, in 1997. Recently, my half sister, Claire, wrote me some healing words about him that will, I hope, serve as a beginning for a closer connection between the two of us: "I often think about how sad and angry Ludwig must have been at the deaths of his parents and friends in the Holocaust. Perhaps the unkind way he treated Jonathan, you, and me was a result of unfinished business with his own parents and guilt that he survived and they did not. I feel that, in many ways, I am also a survivor and you are too."

Claire, her children, my mother, the rest of the Isenberg clan, as well as Rob and his aunt, came together for the first time ever to celebrate Annalise's bat mitzvah and thirteenth birthday on February 18, 2000. As Annalise, surrounded by family and friends, sang the traditional Hebrew prayers in her sweet soprano voice, I felt the sense of belonging and love for which I had searched during much of my life.

Over the years, my life has also been enriched by continuing friendships with many of the climbers with whom I shared these adventures. A few years ago, Joel Bown, Dave Graber, and I finally enjoyed the gourmet feast we'd fantasized during the Endless Winter. For a surprise dessert, Dave baked a luscious cake in the shape of the Parthenon, complete with columns and inch-high frosting.

Fact-checking has served as a happy excuse for getting back in touch with climbing partners with whom I'd lost contact. Wanting everyone who appears in this book to have a chance to comment on it, I summoned my courage and sent the Endless Winter chapters to Toby Wheeler, whom I'd not seen since 1972 in Afghanistan. I was gratified to receive a friendly response containing an apology for his behavior toward me long ago.

One revelatory conversation was with Nancey Goforth, from Bhrigupanth, about our mothers. Like mine, Nancey's mother had received electroshock therapy for depression. The treatments left Nancey's mother unable to adequately care for her four children, and Nancey felt invisible during much of her childhood.

Nancey and I wondered if our family problems might have contributed to our choice—unusual for women—to climb life-threatening mountains. Discussing this possibility, we agreed that we climbed primarily for the beauty, challenge, and companionship. However, we both recalled at perilous moments having the feeling that it didn't matter if we lived or died. We didn't feel suicidal, or have a death wish, or want to get close to death to appreciate life. Rather, we had the sense that as individuals we weren't all that important. We agreed that this feeling contributed to our continuing to climb risky peaks in spite of the losses we had suffered.

Nancey and I acknowledged we could never know for certain the impact of our own childhoods upon our willingness to face death in the mountains when we were younger, but it was interesting to spec-

ulate. We now both have dearly beloved daughters. As mothers, we are no longer willing to attempt dangerous mountains. That and other topics were explored at the Bhrigupanth team's twenty-fifth reunion held in the spring of 2005.

Our Everest team enjoyed our first-ever reunion in October 2006, thirty years after making the second American ascent of Everest. The Annapurna team has celebrated twentieth and twenty-fifth reunions. When we are together, it feels as if we are decades younger, camped on a high glacier with avalanches thundering in the distance. Joan Firey is gone, but everyone else is in good health, and they are still enjoying climbing, professional and volunteer work, and their families.

I have tried to give my daughter a childhood that is 180 degrees different from my own. She first went trekking in the Himalaya when she was six, and attended third grade at the Cloud Forest School in Monteverde, Costa Rica, and eighth grade at Lincoln School in Kathmandu, Nepal. Her father, Rob Gomersall, has remained in her life—and in mine—through letters, phone calls, and visits. He has been living his dream, sailing around the world for the past eight years, and Annalise joins him on his boat annually.

Although Annalise doesn't like the danger or discomfort of rugged wilderness adventure, she has found her own challenges. She has taught English in the Guatemalan highlands, raised money for refugees in the Sudan, coauthored a paper about energy conservation in China, and had an essay published in the bestselling book *50 Ways to Love Your Country* by our good friends at MoveOn.org.

Cloud walking to the summits of our dreams.

My daughter grows up in a new world of opportunities for women. Alison Osius became the first woman president of the American Alpine Club in 1998. Currently, similar numbers of men and women can be found climbing at a high standard on indoor climbing walls and at outdoor areas where ascents involve little risk. However, many more men than women take part in high-altitude mountaineering, the most arduous and hazardous sort of climbing. Nonetheless, before her death climbing Kanchenjunga in 1992, my Polish friend Wanda Rutkiewicz had climbed eight of the fourteen 8,000-meter peaks.

Lynn Hill made the first-ever free climb of the vertical Nose route on Yosemite's El Capitan in four days in 1993 and repeated the feat the next year in twenty-three hours. No one, man or woman, has managed to duplicate her ascent.

Eliza Moran, the first woman president of the International Mountaineering and Climbing Federation, sent me a remarkable report from the 2005 Ouray Ice Festival: "A thousand spectators went crazy when Ines Papert, a young woman from Germany, won the ice climbing competition on an extremely difficult course, beating everyone else by 2 minutes. The doors you opened for others are wide open now." I was moved by Ines's historic achievement and most appreciative of Eliza's kind words.

My climbing and scientific careers have been dormant as I've raised my daughter. Instead, I have been presenting lectures and workshops to a variety of groups ranging from corporate executives to inner-city teens to Mongolian government officials. In my leadership workshops, participants find their own vision of what they want to do in their lives; during intercultural seminars, engineers and managers learn how to work more effectively with their peers in India and other countries.

In 1998, I was invited to a retirement celebration honoring my former Stanford advisor, Buzz Baldwin. Feeling self-conscious about having left scientific research, I nevertheless decided to attend. Peter Kim, a renowned MIT professor, began his presentation by saying that my protein-folding experiments were the basis for current important research on cures for heart disease, cancer, and AIDS. Beneath my surprise and delight at his words, I was left pondering some of the choices I'd made in my life. I had quit doing basic research in the hope of making a practical contribution toward solving global problems; ironically, it appeared that my most fruitful effort toward easing human suffering was my work in protein folding. I was reminded of Buzz's words that although we cannot know the outcomes of our actions, we can listen to a small voice inside us and do what we think is right.

What do I want to do next? I am heartbroken at the destruction of our planet's environment and at the increasing gap between rich and poor. My current plan is to return to my scientific research and public-policy work on toxic chemicals so I can help awaken the public to the dangers these substances pose and to the importance of regulating their use. I continue to believe that if each of us takes slow, steady steps, we can break a trail to the summits of our dreams.

Afterword ▲

The gray November mountains towering above Banff matched my mood as I walked into a packed lecture hall to join a panel on memoir writing at the 2005 Banff Mountain Book Festival. One of the other panelists was the author of the infamous *Outside* magazine article with the cover blurb HAS WOMEN'S CLIMBING FAILED? and the subhead "Why Has the Course of Women's Climbing Led to Tragedy?" After twenty years of avoiding this eloquent critic of women's climbing, I now had to sit within twenty feet of him. My stomach knotted as we exchanged pleasantries.

I began to relax as the panel began and we each spoke about our motivations for writing our memoirs. Then the topic turned to sexism in mountaineering. A distinguished ninety-one-year-old member of the climbing establishment looked straight at me and asked, "Why were male climbers so hostile to your introduction of ladies into climbing? I don't think that happened to Fannie Workman, Miriam O'Brien Underhill, or Annie Peck. Why did you have so much difficulty whereas the other early women pioneers did not?"

I took a deep breath and answered.

"Before our Annapurna expedition, I looked for women's names in a list of climbers who had previously attempted eight-thousand-meter peaks. To my surprise, most of the women who had tried to climb that high had the same name—'and his wife.'

"Miriam Underhill and Fannie Workman were wives of respected American Alpine Club members. One reason my experience was different from theirs is that I am single. Annie Peck, also unmarried, described having problems similar to mine in her books.

"And I'm Jewish—a possible reason I've long refused to believe.

But last week a climbing historian showed me a surprisingly anti-Semitic letter one distinguished climber sent to another."* These climbers were in the same circles of friends as those who had treated me so badly.

The room burst into an uproar. When it calmed down, as though to keep the hullabaloo going the panel leader asked for comments regarding the *Outside* magazine article written by my fellow panelist. The author admitted he had been commissioned by the magazine's editor to write an article showing that women were stepping beyond their abilities in the high mountains and dying as a consequence. He claimed his piece was a complicated and nuanced discussion, and that the inflammatory title and cover line came from the magazine, not him.

I couldn't contain myself. "Your article said that my friends died on Annapurna because I, out of pride for women, refused to let Sherpas accompany them on their summit attempt. That's not true. Your article was not accurate and it hurt me and other women."

As soon as my words were out, the author emphatically disputed them, everyone started talking at once, and I regretted my forthrightness. However, after the panel, people from the audience crowded around me, offering their congratulations for my having dared to confront publicly the twin demons of sexism and anti-Semitism. I felt enormous relief, and then some exhilaration.

Speaking with the elderly climber who had asked why I was treated differently than the climbing wives, I said, "My frank answer came from a place of respect. I wanted to give you my best answer, even if it was upsetting."

He listened, put his arm around me, and said in a kind tone, "Arlene, you know people like and admire you now. It's time for you to forget how you were treated thirty years ago and move on." He seemed sincere and I felt comforted by his words.

---

*The letter described an accomplished climber as "a low-grade New York Jew" and a "mutt" who should therefore not be considered for membership in the American Alpine Club. My thanks to Maurice Isserman and Stewart Weaver for sharing this letter, which is printed in full in their forthcoming book *Fallen Giants: The History of Himalayan Climbing from the Age of Empire to the Age of Extremes.*

For the rest of the weekend, people from the audience continued to tell me that it was important to bring sexism and anti-Semitism into the open. And I felt similar to when, after years of feeling hurt and angry about my aunt Shirley's comment that I'd amount to no good, I finally discussed that childhood memory with my aunt Ruth. Once again, naming an injustice had dissipated its charge. Forgetting my previous pain didn't seem realistic, but I was making peace with my past.

Around the same time, I was also finding closure with my family. During the year that I finished writing *Breaking Trail,* my eighty-eight-year-old mother's life was winding down. I learned to lift her into a wheelchair, feed her, and even change her diaper. But every time I visited her she would brightly suggest that it was time for her to move to California and live with me. And every time I had to disappoint her.

Our favorite shared ritual was going outside so my mother could smoke, a forbidden pleasure inside her nursing home. In the winter, I'd bundle her in a down jacket, woolly hat, gloves, scarf, and lap blankets, reminiscent of outfitting for an expedition. We would wait at the end of the hallway for the world's slowest elevator to arrive. In one hand, I'd balance my mother's glass of water, cigarettes, matches, and extra blanket, and with the other I'd maneuver her into the elevator amid the other wheelchairs, nodding cheerily to the unmoving occupants. Once we reached the ground floor, I'd guide her chair back out of the elevator and into the bitter Chicago winter. Then I would tuck the blankets around her and try to find a quiet corner to light her cigarette. It would take me several attempts, frantically inhaling as the wind whipped around and blew out the match. At last I would coax an orange glow and pass the cigarette to her.

One day when the temperature was below zero and the wind biting, my mother wanted to enjoy her cigarette without going outside. I pushed her wheelchair to the window in her room and opened it, but just after I'd flushed the incriminating evidence down the toilet, we were busted by the outraged head nurse. "You could have burned the building down. Or set off the sprinkler system. You could have caused thousands of dollars in damage."

I said I was the smoker and took the rap.

In the spring when *Breaking Trail* was finished, I showed my mother the pictures in the book and read the family sections to her. Then I held my breath and waited, watching the inert skeletal woman who used to be my tall, robust mother.

"I like it a lot," she finally said with a weak smile.

"Why?" I asked.

"Because it is true," she said. "That's how it was at our house."

By summer, her lymphoma had become leukemia and only blood transfusions kept her alive. The last time I took her outside in the sticky August Chicago heat, she couldn't even smoke. Mostly I held her hand and breathed with her during that visit. But once, she opened her eyes wide and cheerfully told me she was coming to live with me. I couldn't answer her that time.

My mother died a few days later. I brought her ashes back with me to bury in our yard in California, where she always wanted to be.

In the year after *Breaking Trail* was published, I traveled the world talking about the book and reconnecting with friends from the various chapters of my life. Vera Komarkova, who reached the summit of Annapurna I on our expedition, had died of breast cancer in the summer of 2005, at the much-too-young age of sixty-two. But most of the people with whom I shared adventures are still very much alive. I was delighted to see many of them: Fred and Ron from Mt. Adams; Linda from Mt. Waddington; Joel, Tom, Dave, and Annie from the Endless Winter; the Heersinks from Uganda; Jennifer from Kenya; Phil, Barb, Gerry, and Bob from Everest; Irene, Margi, Christy, Piro, and Liz from Annapurna; Susan, Nancey, and Penny from Bhrigupanth.

I also decided to try to meet relatives during my book tour. In Los Angeles, my kindly second cousin Bob Feldman showed me an intriguing picture of our mutual ancestors, the Isenberg clan, formally posed in Davenport, Iowa, around 1917. Bob didn't know who

The Isenberg clan assemble for a family photo in Davenport, Iowa, during the winter of 1917.

the people in the photo were and thought it would be impossible to find out.

However, the following week Sue French, a New York cousin, showed me another family photo, and also a letter from her deceased aunt Edith identifying the people. For the first time, I saw my tall, distinguished great-grandfather, who sold grains and moved the family to Iowa, and my great-grandmother in her babushka—the only literate woman in her shtetl.

Looking again at Bob's photo, Sue and I established the date as sometime in the winter of 1917, when my grandmother was pregnant with my mother. As I examined the image of my grandmother in her maternity dress with the large white collar, I recognized my own face as a girl. That I might in any way resemble my argumentative grandmother was startling. Could my tenacity have come *from* her rather than from opposing her?

Sue also showed me a haunting photo of my grandfather's older brother Sam, the violinist. Looking at his handsome, melancholy face,

Detail of Isenberg family photo. *From left, seated:* my great-grandfather, great-grandmother, and grandmother pregnant with my mother. My grandfather is standing behind my grandmother.

I wondered at the despair that had caused him to take his own life, leaving his wife and five children in the new land of Iowa. I surmised that Sam's suicide might have contributed to my not being allowed to study music, musing on the connection between my great-uncle's depression and the limitations of my childhood—and how, ironically, it was those confines that taught me from an early age to push beyond boundaries, to break my own trails.

Now, thinking of my relatives and friends, both those alive and those no longer with us, I feel part of a much larger and more caring family than I ever imagined while growing up in our small row house in Chicago.

I feel especially privileged to have made many friends in the Himalayas during my years of climbing and trekking. Although the region lacks material wealth and development, it is rich in scenery, spirituality, and community. I am currently passionate about finding

Annalise and me on the Inca Trail in Peru during the summer of 2003.

ways to combat the increase in harmful chemicals in the environment, global poverty, disease, and climate change. The poor of the Himalaya and other parts of the world bear the brunt of all four.

Last year I trekked with my daughter, Annalise, through the golden land of Burma and the Himalayan state of Sikkim. I am helping to support a project bringing water, schools, bridges, and health training to tribal villages in Burma. This summer Annalise and I studied Spanish in Guatemala, a beautiful country with little economic development and a tragically high rate of unemployment; I would like to find a way to encourage investment in that region. I am still climbing mountains, but they are metaphorical rather than physical ones, and the rewards are just as great.

During my years researching and writing this book, I focused on the trails that lie behind me and found a revelatory order and meaning in them. My last months of touring and speaking about the book have further increased my acceptance of my past. Energized by this understanding, I look forward to the future stretching ahead of me like a slope of clean, unbroken snow. I passionately hope my tracks up this slope will lead toward the summit of a healthier environment in a more just and peaceful world. The ascent will be difficult. My route is unknown. But as I begin to break this new trail, moving slowly and steadily upward, I feel joy and confidence that soon all will be revealed.

Acknowledgments ▲

A dedicated community of my climbing partners, scientific colleagues, and loving friends have made enormous contributions to this memoir. I have sometimes found myself a little overwhelmed at their input, but making the changes they suggested always resulted in significant improvements.

My special thanks to my editor and good friend Annie Stine, for her caring and incisive critical and writing guidance; to my agent, Felicia Eth, for her astute counsel and energetic support; and to Lisa Drew, my editor at Scribner, and her able assistant, Samantha Martin, for their good humor, flexibility, and constant encouragement through many travails. Thanks also to Stacia Decker, my talented editor at Harcourt, for her detailed edits, wisdom, and support for the paperback edition.

My gratitude for generous help with photos and maps to the consummate cartographer Stace Wright at Eureka Maps, the artful mapmaker Dee Molenaar, Bradford Washburn for his superlative Denali photos, and my dear friend Martha Geering, the skillful art director at *Sierra* magazine. All authors need an art-director friend.

My loving writing mentor, Luree Miller, first suggested the idea of this book. Over the decades I struggled to write it, I wasn't consistently grateful to her for starting me on the project, but I certainly am now that it's approaching completion. I wish I could show it to her, but Luree died in 1996. Edward Bernstein, my brilliant philosopher lawyer, is another long-term supporter with whom I cannot share this book, as he died in 2004. Luree and Edward's inspiration remains.

For close readings, detailed comments, and wise editing, I owe an invaluable debt to Carolyn Said, Grant Barnes, Laura Waterman, Marissa Moss, Diane Fraser, Penny Kramer, and Lilian Matimore.

B. K. Moran, Terry Byrnes, Sylvia Margolin Paull, Rob Gomersall, Howard Simon, Martha Platt, Marcus Moench, Joel Bown, Ignacio Tinoco, Diane Bernbaum, Betsy White, and Anne Messer gave me early and continued support and spurred me on with their enthusiasm.

I was privileged to share the beauties and dangers of the mountains with many companions who also were kind enough to contribute to this project: Fred Hartline, Tom Dunne, Paul Pennington, Dick Birnie, Louise Wholey, Tom Stephens, Bill Dimpfl, Les Wilson, Linda Crabtree Nowell, Margaret Clark, Dana Isherwood, Mike Bialos, Dave Graber, Toby Wheeler, Dave and Annie George, Hans Brüyntjes, Molly Higgins, Heidi Lüdi, Dan Emmett, Bob Cormack, Barbara Roach, Gerry Roach, Johan Reinhard, Richard Isherwood, Irene Beardsley, Margi Rusmore, Dyanna Taylor, Liz Klobusicky, Piro Kramar, Christy Tews, Penny Brothers, Nancey Goforth, Susan Coons, and Mike Fahmie.

For helpful feedback on the entire manuscript, my sincere thanks to Steve Roper, George Cummings, Louis Reichardt, Nick Clinch, Eliza Moran, Maurice Isserman, Nicole Gillespie, Ann Manheimer, Marci Rubin, Marianne Betterly-Kohn, Melody Ermachild Chavis, Marcy Berkman, Annie Tiberio, Gary Marcos, Ann Stein, Lisa Bellm, Bruce and Judy Todd, Carl Shapiro, Jill Cherneff, Rochelle Blumenfield, Santosh Philip, Barbara Banks, Pat Gregory, Debra Denker, Hilary Goldstine, Lynette Levy, Eileen Kramer, Cathy Luchetti, Sandy Curtis, Linda Liscom, Xenia Lisanevich, Ellen Lapham, Carla Dole, Ted Vaill, and Ella Ellis.

For valuable critiques on portions of the manuscript, my gratitude to Eric Hansen, Joan Blades, Carol Kornblith, Jill Vialet, Kirsty Melville, Julie Van Gelder, Caroline Pincus, Beverly Hartline, Claire Schoen, Elisabeth Caspari, Lesley Quinn, Wendy Lichtman, Peggy Vincent, Joan Lester, Penny Kramer, Tanya Atwater, Libby James, and Sharon Kinsman.

Melissa Bloom, Alyssa Johl, Christine Jackson, Ann Hester, Heidi Botts, T. J. Roethe, Lisa Nelson, and Mary Pronk helped in innumerable ways: picking up the many pieces I couldn't manage to juggle, staying cheerful, and finding all the things I lost.

For sharing information and other important contributions, my sincere thanks to Annie and Orton Hall, Bill and Kris Hall, Barbara and David Carson, Charlie and Mary Lou Swift, Myra Lazerwith Gervutz, John Harlin III, Vern Tejas, John Mock, Juliana Roth Ordoñez, Rabbi Jack Gabriel, Marleen Kleijnen, Ruth Schneebeli-Graf, Roger Wise, and Judy Kunofsky.

For comradeship in the laboratory and commentary on the scientific part of my life: Bruce Ames, Buzz Baldwin, Jean-Renaud Garel, Scott Emmons, James Huston, Barbara Dengler, and Mark Levine.

Marti Hearst; Stans Kleijnen and Brian, Simone, and Joop Verbaken; Roshan Bhajracharya and family; Bruce, Sophie, Conrad, Maeanna, and Cynthia Welti; and Raphael Shannon kept my morale high by providing me with a place to write, a warm meal, and an open ear.

My appreciation for their hospitality and kindness also to the gracious people in the many countries where I was privileged to climb and trek. These adventures wouldn't have been possible without the unstinting support of our staff, porters, and local residents too numerous to name.

Much of the first draft was written from 1995 to 1996 when Annalise and I lived in Monteverde, Costa Rica. Many members of the community there provided us with extraordinary hospitality and assistance. In particular, I'd like to thank Sue Trostle, Mary Newswanger, Debbie Rockwell, Eston and Mary Rockwell, Katy Van Deusen, and Joan Gregerson for their comments on the nascent manuscript, and Stella for her blackberry pie, an ingredient of inspiration.

My gratitude to my relatives for reading the manuscript and being comfortable with sharing some family secrets: Mike and Nancy Vogler, Steve Vogler, Ron Meyer, Claire Blum DeStephens, Jonathan Blum, Jimmy Isenberg, Bob Feldman, Sue French, Cele Rego, Marlyne Weiner, Marjorie Cooper, and Jim Crane.

Annalise and our cats, Micou and Midnight, gave me constant loving company, with somewhat limited patience for late meals and other effects of my focus on writing.

Using my diaries, articles written following my trips, and my imperfect memory, I have tried to describe my experiences honestly and accurately. Whenever possible, I verified information with teammates, colleagues, and friends. However, our recollections are colored by our personal perspectives, and I recognize that others might have different memories of some of these events.

The childhood vignettes in this book occurred during my first seventeen years; the climbs during another period of similar length. The writing, spanning twenty-four years, would never have been completed without the extraordinary love, support, and encouragement of the people I list above, and many others. My heartfelt thanks.

Index ▲

Abbreviations: f.a. first ascent; f.w.a. first women's ascent

About the Author ▲

Arlene Blum is a writer, mountaineer, and biochemist who has taken a leading role in more than thirty successful mountain expeditions, including climbs of Annapurna I, Everest, and Denali. Blum has a doctorate in biophysical chemistry and has taught at Stanford, Wellesley, and the University of California, Berkeley. She lives in Berkeley, California, with her daughter, Annalise, and is currently working to forge a partnership among industries, scientists, and environmentalists to reduce harmful chemicals in people and the environment. Her Web site, www.arleneblum.com, includes color slide shows illustrating the chapters of this book.